D0931486

THE DYNAMICS OF SCIENCE AND TECHNOLOGY

SOCIOLOGY OF THE SCIENCES

A YEARBOOK

VOLUME II – 1978

THE DYNAMICS OF SCIENCE AND TECHNOLOGY

*Social Values, Technical Norms and
Scientific Criteria in the
Development of Knowledge*

Edited by

WOLFGANG KROHN, EDWIN T. LAYTON, JR.

and

PETER WEINGART

D. REIDEL PUBLISHING COMPANY

DORDRECHT:HOLLAND/BOSTON:U.S.A.

Library of Congress Cataloging in Publication Data

Main entry under title:

The Dynamics of science and technology.

    (Sociology of the sciences ; v. 2)
    Includes bibliographies and index.
    1. Science – Philosophy. 2. Science – Social aspects.
3. Technology – Philosophy. 4. Technology – Social
aspects. I. Krohn, Wolfgang, 1941–   II. Layton, Edwin
T. III. Weingart, Peter. IV. Series.
Q175.D94   500   78–1120
ISBN 90–277–0880–0
ISBN 90–277–0881–9 pbk.

Published by D. Reidel Publishing Company,
P.O. Box 17, Dordrecht, Holland

Sold and distributed in the U.S.A., Canada, and Mexico
by D. Reidel Publishing Company, Inc.
Lincoln Building, 160 Old Derby Street, Hingham,
Mass. 02043, U.S.A.

Printed in The Netherlands

TABLE OF CONTENTS

INTRODUCTION

The interrelations of science and technology as an object of study seem to have drawn the attention of a number of disciplines: the history of both science and technology, sociology, economics and economic history, and even the philosophy of science. The question that comes to mind is whether the phenomenon itself is new or if advances in the disciplines involved account for this novel interest, or, in fact, if both are interconnected. When the editors set out to plan this volume, their more or less explicit conviction was that the relationship of science and technology did reveal a new configuration and that the disciplines concerned with its analysis failed at least in part to deal with the change because of conceptual and methodological preconceptions.

To say this does not imply a verdict on the insufficiency of one and the superiority of any other one disciplinary approach. Rather, the situation is much more complex. In economics, for example, the interest in the relationship between science and technology is deeply influenced by the theoretical problem of accounting for the factors of economic growth. The primary concern is with technology and the problem is whether the market induces technological advances or whether they induce new demands that explain the subsequent diffusion of new technologies. Science is generally considered to be an exogenous factor not directly subject to market forces and, therefore, appears to be of no interest. As any case study in the history of technology that includes considerations of the cognitive preconditions of technological advances will show, such a picture is a drastic simplification. A careful analysis of the genesis of a new technology, such as those of the radio and the turbine included in this volume, reveals a complex interaction of a series of factors: preceding developments in the sciences, a specific configuration of economic possibilities and constraints, different orientations of scientists, engineers and entrepreneurs which make them recognize the practical potential of certain knowledge or prevent them from doing so, etc. Indeed, the number and variety of factors is so complex and their conceptualization so difficult that generalizations from case studies seem impossible at this point.

If we look at the history of technology we often find an over-emphasis on the autonomy of technological development and likewise, a stress on the ingenuity of individuals, on the 'heroism' of the lone inventor. Again, this view, although somewhat supported by detailed technical analysis, usually fails to convey the social context out of which the determinants arise that explain the behavior of the individual protagonists of technological development, and account for regularities as well as differences in time and geographical location.

If the historian of technology were to look to sociology for help, he may find clues in the general theory of society and social action, but he will be left with the task of having to 'translate' that theory in order to be able to apply it to the case of technology. Even though technology is one of the most important forces that shape modern societies, sociology has paid surprisingly little attention to it. And the little interest that is shown is focused primarily on the consequences of technological change, while a coherent explanation of the social forces that bring it about is lacking. The sociology of science has made a neat conceptual differentiation between science and technology relegating the latter to the sphere of the production and use of knowledge mediated by social interests. While in principle this is a fruitful perspective, it is limited to technology alone. Science, on the other hand, is perceived to be fundamentally different and the question posed is not which social forces shape its contents but which (functional) prerequisites have to exist in a society to make it possible as it is. Thus, the relationship between science and technology is an ephemeral problem, if any at all, just as technological development itself.

This negligence of the patterns of intellectual development on the part of sociology which, to some extent, can also be attributed to the sociologist's lack of detailed knowledge and his emphasis on social structures could well be expected to be compensated for by the history of science. The history of science and from an epistemological viewpoint, also the philosophy of science attempt to decribe and/or explain scientific development as a cumulative and progressive evolution of ideas where, at least in principle, each step in that process can be shown to be a consequence of the preceding one. Similarly, as in the history of technology, it provides 'internal' accounts of the evolution of scientific thought proceeding on the assumption that the social and economic context in which these ideas emerge plays at best a trivial role in that it makes science possible or impossible. Technology, on the other hand, is seen as a mere 'spin-off' from scientific ideas, alien to science exactly because it is subject to social needs and interests.

From this hopefully fair description of the respective foci and blind spots, analytical strengths and weaknesses of the disciplines that are concerned with the study of science and technology, it is evident that the problem of the relationship between science and technology falls between the lines of disciplinary demarcation. This is not, of course, a new insight, but little followed from it as is exemplified by the lack of response to Lewis Mumford's *Technics and Civilization*. Thus in 1959 Mumford who very early had tried to transgress the lines of disciplinary perspectives was invited by 'Daedalus' to review his own book. When this book originally appeared in 1934, Mumford recalled 'it stood alone in its field'. 25 years later, he felt compelled to comment "Whatever the original defects of 'Technics and Civilization', whatever shortcomings time has disclosed, it still unfortunately possesses its original distinction: It stands alone, an ironic monument if not an active influence".

The answer to the question formulated at the outset as to the causes of the growing interest in the problem is that in all likelihood both the relationship of science to technology is changing and the awareness of it is growing. Indications of the former are new types of institutions such as problem orientated research installations in which the activity of knowledge production transcends the borderlines between basic research and technology, an increasing convergence of the character of scientific and technical training and of the professional organization emerging from a similar convergence of methods, procedures and bases of knowledge. All of these used to be accepted criteria of differentiation between science and technology, but they do no longer seem to be fit for that purpose.

The volume we have compiled cannot pretend to trace all these changes and/or to do this by approaching the phenomenon with an integrated inter-disciplinary analysis that avoids all the shortcomings of the established fields. To formulate programmatic standards is one thing, to realize them in concrete research is quite another and if the problematic of research on science and technology did not exist, obviously there would be no need for a new programme. It is not even clear if it is realistic to expect that the disciplines in question will eventually converge in an integrated approach in the study of the relationship of science and technology as the complexity of such an enterprise would be immense and the advantage of controlling different variables for different goals of analysis would be lost.

Aside from such methodological caveats the compilation of articles that are supposed to be focused on and in consonance with a new and ambitious programme is to a considerable extent a matter of risk and chance. As often happens, the final product looks somewhat different from what was originally planned. It would demand continuous research collaboration to achieve consensus in outlook and thus homogeneity in their writings among a group of authors when attacking a complex issue such as the one presented here. This is what this volume may contribute to, it was not the context from which it emerged. This collection of essays may be said to represent a state of research, however, which from various disciplinary viewpoints, provides the starting point for the kind of analyses we have in mind. Thus, although the individual contributions need the reader to relate them to the research programme, we hope that as a collection they are compelling enough to elicit discussions among historians of science and technology, economic and social historians as well as sociologists.

W. KROHN
E. T. LAYTON, JR.
P. WEINGART

AFFILIATIONS OF AUTHORS

Prof. P. Janich, *Universität Konstanz*
Dr Ron Johnston, *Manchester University*
Dr Tom Jagtenberg, *School of Sociology, University of Wollongong*
Prof. Edwin T. Layton, Jr.,*University of Minnesota*
Prof. Hugh G. J. Aitken, *Amherst College*
Dr G. Küppers, *Universität Bielefeld*
Prof. Karl-Heinz Manegold, *Technische Universität, Hannover*
Judy Sadler, M. Sc., *University of Manchester.*
Prof. Patrick Fridenson, *Université Paris X – Nanterre*
Dr Gernot Böhme, *Technische Hochschule, Darmstadt*
Dr Wolfgang van den Daele, *Max-Planck-Institut zur Erforschung der Lebensbedingungen der wissenschaftlich-technischen Welt, Starnberg*
Dr Wolfgang Krohn, *Max-Planck-Institut zur Erforschung der Lebensbedingungen der wissenschaftlich-technischen Welt, Starnberg*
Prof. Peter Weingart, *Universität Bielefeld*

PART I

SCIENCE AND TECHNOLOGY
The Conceptual Distinction Reconsidered

PHYSICS — NATURAL SCIENCE
OR TECHNOLOGY?

PETER JANICH

Konstanz

Modern physicists understand themselves in a *naturalistic* way. Critical questions which philosophers of science ask about reasons for the development of long term research, about the theoretical structure of the results of such research or about the relation between fundamental research and application, usually end with remarks about constraints called, for short, 'natural laws'. These answers are given despite some facts which relate to such different matters as what physicists really do when they engage in reasearch, what conditions on the institutional level physical research must conform to, and what traditions of investigation into nature they follow.

This paper criticizes this understanding of physics by physicists. It proposes an alternative classification of science and technology, and tries to exhibit methodological and political consequences of such an alternative understanding. Thus, in proposing a *critique of science* the paper makes use of a philosophy of science which does not remain stuck at the metalevel of description as if it were to satisfy an abstract need for philosophical classification (which itself seems to require a freedom from concrete consequences). This critique of the physicists' self-understanding, therefore, does not share with such a self-understanding of science the view that critical reflection is part and parcel of an ideology divorced from science's every day practice. What characterizes the claim of this paper is this: beginning with a 'theory of science' which since the Vienna Circle has become an independent discipline, the analysis here should lead to a kind of philosophy of science where the word 'philosophy' stands for the claim of ancient philosophers to be engaged in the search for that kind of knowledge which leads to better action.

Krohn/Layton/Weingart (eds.), The Dynamics of Science and Technology.
Sociology of the Sciences, Volume II, 1978. 3–27. All Rights Reserved.
Copyright © 1978 by D. Reidel Publishing Company, Dordrecht, Holland.

1. What Does it Mean to Say that Physicists Understand Physics in a Naturalistic Way?

The roots of the words *physics* and *naturalistic*, one Greek and one Latin, refer to 'being born', that is, to organic, living parts of *nature*. Physics, however, owes its enormous 'success', to the accomplishments of seventeenth-century physicists in abandoning the Aristotelian approach. The core of this approach was an analysis of motion in the guise of a theory of causes. This analysis, on the one hand, could specify the motions of free fall and projection as a special class of motions explainable by exterior causes as opposed to another class of motions, organic ones, which called for a different explanation. But on the other hand, this analysis could not reach the successful idea of classical mechanics, namely the principle of inertia. Nature, in the specific sense of classical physics and despite the literal sense of the words 'physics' and 'nature', is the inanimate. Consequently from its beginnings in the seventeenth century until today physics is better referred to as *mechanics*. For, more or less burdened with theological ideas, the world as the object of the physicist's research is to be seen as a large machine the functions of which require investigation, description and exploration.

The statement that physicists understand their discipline in a naturalistic way then always refers to a terminological shift of 'nature' from the organic or living to the mechanical or technical.

General statements such as 'physicists understand their discipline in a naturalistic way' are difficult to argue for by means of physicists' explicit declarations. Only rarely do physicists make judgements about their discipline as a whole. But sometimes in physics textbooks belonging to an older tradition remarks can be found on the purpose, the object and the nature of physics. Statements occur here about the need to investigate 'nature' or 'natural phenomena' especially 'an immense multiplicity of natural phenomena'. In every day life situations regularities would have already imposed themselves on the man in the street. Observations of those regularities moreover would have already lead, especially since the introduction of mathematical methods, to the successful discovery of natural regularities, in short 'natural laws'. Despite the fact that a highly developed set of instruments, both of a conceptual and nonlinguistic kind, has been introduced into physical research, it is said that the physicist's curiosity is directed to natural phenomena by technical means which are at best isolated in a pure form or demonstrated as dependent on completely manageable conditions. There is also much talk about the seeking and finding of natural laws, a metaphor which suggests that natural phenomena are searched for like rare plants or minerals. A list of such metaphors would include among others: that the book of nature is written in mathematical symbols, that the universe shows itself to be ordered or to be a cosmos in the original Greek sense of this word, that a deeper penetration of the secrets of nature is the just reward for patient search after the laws of nature, and so on. All such metaphors show that 'nature' defined as the object of inquiry is what it is because there exists an order in nature which becomes all the more manifest the more cleverness is invested in research.

Thus put, physics remain part of a *mythical tradition*. No modern physicists would like to personalize nature. Nature is *objectified*. It is defined as a given object about which empirical research can teach us that certain laws are valid in it. Even the major crisis in the foundations of physics, which lead to the relativity theories on the one hand and to quantum physics on the other hand, could not overcome this conviction. Despite the often repeated opinion that relativistic theorems about space and time are not to be interpreted in a realistic way as H. A. Lorentz did but in a conventionalistic way as A. Einstein did, the relativistic theory of space and time still holds way among physicists as a set of natural laws. And despite Bohr's insight into the complementarity of the given and the produced in microphysical observation, i.e., despite the insight that only the whole complex of the given and the disturbance of the given by observation is open to physical description, this basic problem of micro-physics is interpreted as a particular difficulty about the *access* to a certain range of magnitudes corresponding to analogous problems in macro-physics. But the reasons for these difficulties of access are looked for once again in the field of natural laws.

One could, of course, suppose (and support it with good reasons, e.g., that the self-understanding of physicists are always something *post factum*) that the opinions of physicists about physics is part of a dilettante philosophy in the articles of older prominent physicists, or an ideological phenomenon of the 'Überbau' presented in the superficial introductions to textbooks. In either case such opinions could be judged as not seriously defendable and also as practically irrelevant, lacking any real consequences. These meta-statements do not, one could suppose, say anything about relevant features of the self-understanding of physicists and do not have any bearing for physics at all. Such a supposition, however, would be wrong. There are methodological and political reasons to assume that the physicists' naturalistic misinterpretation of their discipline entails *concrete consequences*.

Although the acknowledged theories of modern physics stem from such different sources as astronomy, caloric theory, theories of machines and of the magnetism of the earth, optics, and so on, physicists nevertheless assume that they are dealing with *one nature* which, therefore, has to be described by *one theory*. If historically there occur partial theories which are logically inconsistent with each other, at least one of these partial theories must be wrong. The unification of historically developed theories into one big theory which covers all natural phenomena (and, in the ideal case, is an axiomatic theory) has been, since the beginning of the nineteenth century at least, a naturalistically based aim of physics.

A naturalistic understanding of the *history of physics* corresponds to the naturalistic understanding of theories like those mentioned above. Thus, the history of physics is written as a chronology of successive discovery. As far as theories are concerned, history leads to a growth in the unification of knowledge by *imbedding* partial theories into more comprehensive ones. For example consider the imbedding of acoustics into mechanics or of optics into electrodynamics. Naturalistically, the possibility of arranging theories in this way is not just simply stated, but it is assumed that physicists are forced to act this way by natural laws. Acoustic phenomena are said, for example, *to be* mechanical ones, optical phenomena *to be* electrodynamical ones. But the methodological decision to explain acoustic phenomena by means of mechanics or optical phenomena by means of electrodynamics is not seen.

Part of this idea of imbedding partial theories into more comprehensive ones is the belief that older theories in physics are not falsified but defined more precisely with respect to the domain for which they hold. For instance, the relation between the Galileo-invariant mechanics and the

Lorentz-invariant electrodynamics is interpreted in such a way that classical mechanics still holds for the domain of small velocities (relative to the velocity of light).

Nature and its laws are still held to be the most important basis for the historical development of theories. Even in the extreme case (which has already occurred in connection with relativistic physics) that a theory, interpreted as a true or corroborated picture of nature or its laws, makes statements about instruments of measurement which differ from those statements which led to the construction and the use of these instruments, physicists prefer to classify these instruments as natural objects and interpret the principle of construction as a kind of preliminary error on the every day level. Although they owe their knowledge about nature to the artificial properties of their measuring instruments, they interpret these instruments as approximations of natural laws (e.g., the euclidean properties of instruments used to measure length as an approximation to the non-euclidean 'structure of space'). As long as the principles guiding the construction of instruments can be interpreted consistently with the empirical theorems based on measurements made by these instruments, no further problems are raised.

Finally the *scientific* character of physics, despite the discussions among philosophers of science over three generations, is construed as the *objectivity* of physics. And objectivity again is interpreted as independence from space and time. It is, therefore, located in the domain of natural laws since independence from space and time is a special mathematical property of certain physical theorems held to be empirical. As a consequence, problems which arise between natural scientists and humanities specialists in a university's every day practice were usually explained among physicists with the argument that the object, 'nature', of their own discipline implies objectivity, whereas the object of the social sciences and the humanities, namely human actions and their results, must lead to endless debates and unresolvable conflicts between schools. Here the methodological characteristics of naturalistic self-understanding merge into political ones. It is held to be a question of scientific empirical knowledge only as to whether a comprehensive theory about nature is possible, research which is independent of any political or moral discussion about its own goals is suggestive. Therefore it is not surprising that those who defend the principle of research as *value-free* ('wertfrei') are mainly natural scientists. Here the value-free character of physics is commonly understood as neutrality with respect to application. The ancient philosophical view that any kind of knowledge is valuable just because it finally enables men to act better is reduced to a mere defensive strategy against questions asked of physicists concerning the goal or the purpose of fundamental research in physics. A largely subjectivistic and hedonistic idea of being a scientific investigator, of making important discoveries, and of being free of all responsibilities for bad applications of their own results corresponds to the wide-spread distinction between fundamental and applied research.

Although it might be easy to defend this distinction as a reasonable application of the principle of the division of labour, the distinction is more usually confirmed by naturalistic arguments: whereas applied research in the defined sense of research whose results are addressed to producers and consumers, is directed by needs or wishes of these clients, basic research addresses other scientists and serves to reveal the truth about nature. Only this program makes a belief in the inner laws and independent destiny of basic research plausible.

2. Nature and Technology

The naturalistic understanding of physics can be criticized from several different points of view. One of the best developed approaches is centered on theories of scientific languages. This approach points out that on the one hand any science is necessarily bound to represent its results linguistically, and on the other hand any terminology, that is to say any system of linguistically communicable distinctions, has a conventional component which is not determined by nature or its laws. Another well known criticism is based on the weakness of the naturalistic interpretation of the history of physics. The concept of nature can be proven to have changed so basically during the history of culture – and to have changed much because of science – that it would be absurd to speak of nature as an historically enduring and identical object of science. In particular, the relation between man and nature in the areas of food production and manufacturing of goods disallows any assumption that there is a nature which can be investigated independent of all historical presuppositions.

Here, however, we shall follow another approach. An unbiased consideration of physics shows that physicists have a great deal to do with *instruments* – an observation which does justice to the unquestionable fact that physicists as empirical scientists gain their experience only by the use of particular instruments, apparatuses and tools. The experience of modern natural science is *apparative* experience, and experience necessarily structured by instrumentation. The instruments used for the collection of empirical results in the form of protocols of observations or measurements are themselves constitutive of these results. Closer consideration allows us to distinguish at least *three kinds of physical instruments* according either to the intentions which are pursued with each, or dependent on the presuppositions of their construction or use.

One class of instruments is used for the *observation of natural phenomena*. These are, in the first place, visual observation of astronomical phenomena, the naturality of which consists in their being independent of men. This independence of planets and stars and indeed of any meteorological or geological phenomenon means that it 'is there', that it can be seen without instruments in at least a diffuse form, and above all that it is not an artefact, i.e., not produced by man. To be sure, the time has passed when it could be a scientific discovery of first rank to explain by means of a telescope the spots on the surface of the moon as shadows of craters. Today physicists are also accustomed to say that natural phenomena are 'described in a quantitative way'. In other words, what shows itself, shows itself as a 'numeric datum' relevant for a theory and relevant only in connection with a particular measuring tool. But this view no longer distinguishes between results of measurements and results of observations, although any use of a measuring tool always makes a 'factum' out of a 'datum' merely by the conventional choice of scales or units.

Here *measuring tools* count as a second class of physical instruments. It does not make sense to say that they are used for the observation of natural phenomena. Although it is indisputable that measurements say something about what is measured, measuring instruments serve to produce artificial phenomena. A clock very well exemplifies this, and it is important to realize the difference between a clock and, say, a microscope. A microscope allows one to observe the inner structure of an organic cell.

This cell is an object which exists independent of the fact of its observation. Clocks however do not show a 'natural object' time or the passage of time as if this object were something 'natural' like the flow of the river Rhine. Time, or more precisely, the domain of temporal statements relevant to physics, does not naturally exist. It is, rather, a *cultural product* which allows communication about comparisons of events with respect to their duration or succession. These comparisons are then generalizable by being referred to a standard motion which is artificially produced as a uniform motion of the clock's hand. Since the days of classical mechanics, uniform motion has been an imitation of the earth's rotations, a fact which led the technology of previous times to divide the day for purposes of organization or for special problems of medicine (pulsilogium), astronomy or navigation. Only after Kant's arguments against the earth's rotation as an ideal standard for the measurement of time is it possible to interpret the history and the improvement of clock-making as guided by a norm, even if not explicitly formulated, which prescribes a uniform motion independent of a natural example.

Therefore, the *duration* of an event which a clock measures is an *artificial phenomenon*. And the situation of other measuring tools is analogous. Only if one were to forget all the actions necessary for the non-verbal production of a measuring rod and all the verbal conventions as well could one say that a natural object such as a broken branch 'has a natural length'. 'Length' can be spoken of in the sense of a length ratio between two objects, and a length ratio can be defined operationally only for straight connections between pairs of points. Being *straight* is a property of the edge of a stick or of a ruler and therefore has to be achieved artificially, that is, by a technology. In other words, the production of straight edges precedes the possibility of stating that a particular natural object has a particular length. This is especially so in a scientific context where explicitness is required. In such a context even a system of *prescriptions* about how to produce straight edges without the use of other straight edges or measurements of length is methodologically prior to the measurements themselves (1).

In summary, whereas natural phenomena like clouds, stars and lakes, can be observed because they exist independent of any human action, lengths, durations, velocities, accelerations, masses, forces, frictions, charges, tensions, etc. do not exist independently of human actions. Such entities are produced as quantified qualities by technical instruments like rigid rods (and their technical refinements), clocks, balances, and so on.

If the two classes of instruments mentioned up to now are characterized in terms of their purpose, that is, *making observations and measurements technically possible*, the range of instruments which physicists actually use is not yet exhausted. To see this, one needs to regard an arbitrary experiment. Rather experiments consist of artificial situations in which processes are initiated and then left to their own running. Experiments should lead to statements of the form: 'if a situation, *s* (as defined by a set of conditions described in physical parameters) is produced, then the process, *p*, occurs (which includes the special case that a "process" *p* is a state)'. To be sure, measuring tools and possibly observation tools are required to control *s* and *p*, but apparently the experimenter needs more devices than just these. A simple historical example can clarify this point. To 'test' the Galilean law of free fall experimentally, certain conditions of the experiment in addition to the necessity of measuring length and duration with some measuring tools must be technically realized. The experiment requires for instance an inclined plane, a smooth ball or cylinder, and a vacuum chamber to avoid air friction. To be sure the properties of the experimental device must be controlled by measuring tools; but in addition the *experiment* itself requires its own *device* (and would require it even if the conditions were not controlled quantitatively). This simple example, which on this point does not differ from the most sophisticated modern experiments, can show that doing experiments is more an activity to produce *technical effects*, which can be described appropriately as engineering rather than as a scientific activity, properly speaking, as a construction of machines rather than as an inquiry into nature, as an attempt to produce artificial processes or states rather than as a search for true sentences. The actions of an experimenter would also make sense if the interest in the momentary technical effect completely superseded the interest in learning something from the experiment for later repetition.

The physicist therefore, in order to be able to learn by his successes or failures, needs at least a conjecture about the function of his experimental device, i.e., he needs a criterion for the *technical* success of his experiment as opposed to a criterion which would concern a theory to be tested by the experiment. And because the physicist 'suggests' something about what he himself has produced, or at least designed, it is more appropriate to say that, on the level of his technical knowledge, he requires a *plan* for his experiment. The counterposition to the naturalistic interpretation of experiments therefore is to assume that physical experiments consist of technical production of machines according to plans some of which are technically successful and some of which are not.

A closer look at the three classes of instruments used by physicists enables us to criticize a belief which is widely spread not only among physicists, but also among philosophers of science. According to this view, all knowledge of nature begins with observation provoked by natural regularities and leads to science by the gradual introduction of technical observation devices which function like artificial limbs in the following way. The observation instruments extend the range of our sense perception (e.g., make available frequencies not accessible to our eyes or ears) and they objectify sense perception by ruling out subjective standards for sensual qualities. Theories about scales of different levels are often added here which try to represent the introduction of technical aids to observation as a construction running through a scale of scales until in the best case the metrical scale is finally reached. Accordingly, experiments on this view are interpreted as merely additional technical attempts to realize natural events under clearer and purer conditions by successively sampling out particular parameters as constant or variable.

This opinion, however, cannot do justice to the role which the artificial properties and their intentional presuppositions play in the empirical results. A more unbiased attempt to understand what physicists really do, however, can show that no explanation of how physical theories depend on experience is satisfactory unless the constitutive intentions of inventors, builders and users of instruments are taken into account. Consequently, the central *task of a philosophy of physics* is taken here as the task of clarifying in what sense sentences found in physics textbooks, and often called 'natural laws' by physicists and others, really are based on '*a* nature' outside of human actions in the form of a technology, and what parts or components of physical theories are descriptions of plans for technical practice, i.e., of activities which if put into the form of general statements are based on norms rather than on experience.

3. Theory of Physics as Technology

The following section of the paper develops a sketch of a philosophy of physics which owes its main features to the insight that physical experience is always experience gained by means of instruments and therefore correctly understandable only in terms of human actions. In some parts the sketch refers to elaborated non-empiristic theories especially with regard to measurement (2), whereas in other parts the sketch assumes the form of a program (3). The following points are central to this sketch:

(1) The 'naturalistic' understanding of physics will be replaced by an alternative view which may be termed as the 'constructive' view of physics.

(2) Consequently, the relation between natural science and technology will have to be defined in a new way. Natural science is to be understood as a secondary consequence of technology rather than technology as an application of natural science.

(3) This revised understanding of physics as technology will not only have methodological implications which are important for contemporary philosophical discussion but will also have consequences on the principles used for evaluating the history of physics and for actual policy formation in the field of science research programming and of science teaching as well.

However, this sketch does not include a discussion of its own relation to an analytic or empiricist philosophy of science.

(a) Observation

The objections against naturalism presented above can be summarized as follows: Except for some 'applied' parts of physics (i.e., astronomy, meteorology, geography, etc.), physical experience does not consist of a technologically supported observation of nature. Although it is common in the present discussion among philosophers of science to criticize older positions concerning the concepts of observation and the observer (a criticism which is directed against particular empiricist programs of reductionism), lectures and textbooks of physics still emphasize the role of 'the observer'. This concept was successfully introduced into physical discussion mainly by A. Einstein's proposals to explain operationalistically the relativistic concept of time by means of observers who are moved relative to each other. Independently, the Bohr–Einstein discussion in the beginning of quantum physics, and the Bohr complementarity between the observation and the observed focused the attention of physicists more sharply on the role of observers. But neither issue, the relativistic nor the quantum physical one, construed the 'observer' in the sense of a human being who acts purposefully and tries to choose appropriate means. Rather, both reduced the observer to an abstract entity which registers physical parameters. So the use of the words 'observer' or 'observation' in physical discussions entails a terminological inconsistency which is first to be avoided.

To extend the every-day-use of the expression 'to observe something' to the activity of physicists collecting nonquantitative or quantitative data by means of instruments fails to take into account the difference in *presuppositions*. It suggests that measurements as a precise form of observation in physics is mainly sense perception just as every-day instances of observation are mainly sense perception. This, however, is wrong although nobody would deny that sense perception is indispensable in scientific observations too.

Terminologically, *measurement* and *scientific observation* can be distinguished in this way: observations presuppose *empirical* theories of the observation instruments, whereas measurements do not depend on empirical theories (where 'empirical' again means being based on measurement). The argument is not directed against the obvious fact that in both cases, in scientific observation and in measurement, observation in the everyday sense is involved.

The terminological proposal to distinguish scientific observation from measurement by determining whether empirical or non-empirical theories are presupposed is not arbitrary but corresponds to the historical roots of both activities. Both, observation of nature (in the sense of the non-artificial environment of man) and measurement are much older than the beginnings of science in ancient Greece (which is the beginning of science as an activity aiming at theories). But whereas *observation of nature* – of course, without instruments – goes back to any prehistoric time for which we assume the existence of men, *measurement* is a *cultural activity of men in society* undertaken with the aim of managing problems of organization. Examples are land distribution, comparisons of weight, length, surface or volume of goods, and the introduction of tax systems. Whereas the task of observation was producing knowledge of and orientation within 'nature', measurement served to technically improve certain social interactions. Measurement as an organizational means of regulating these interactions is a much older phenomenon than measurement as a basic art for technologies like, for example, architecture. Only within the reductionistic misunderstanding in empiristic philosophy is physics based on observations called measurements and thus 'facts' as results of observations become confounded with 'facts' as a result of measurements. If 'facts' within physics have something to do with experiments at all, then the experimental conditions must be technically reproducible. And in order to determine whether the technical reproduction is successful measuring instruments are needed. And the relevant properties of these instruments are technically reproduced themselves.

(b) Measurement

The naturalistic view that measurements are highly stylized observations led to *empiricist theories of metricization* (H. v. Helmholtz, R. Carnap, C. G. Hempel *et al.*). The basic idea of these theories consists in running through a scale of scales by stepwise differentiation beginning with simple predications, going over certain relations, ratio scales and so on. The highest level of scales, the metric scale, is of special interest for physicists. The metric concepts specified by this highest level not only have the richest logical structure but they are the only ones which require a definition of *units*. Besides the difficulty that the sentences characterizing the logical structure (e.g., the transitivity of order and equality) cannot be empirical because their empirical control would presuppose their own validity (4), the empiricist proposals for defining the units are in addition logically circular. The knowledge necessary for the definition of a reproducable unit (of length, duration, mass, charge, etc.) is accessible only by means of measurement.

However, the empiricist theories of metricization, which are properly called naturalistic theories because they assume natural laws as conditions for the reproducibility of units, interpret the art of measuring as the art of finding reproducible units. These approaches therefore overlook the fact that, although human efforts to reproduce technically certain properties of measuring instruments can fail because of the natural properties of the materials in use, nevertheless the failure can be stated only in relation to the intentions and capabilities of acting men. To provide *reproducibility* is therefore primarily a matter of the *organization of actions in such a way that their repetition leads to reproduction of instruments*.

Setting aside the empiricist philosophy of science, consideration of the practice of research in physical laboratories shows that those productive acts do not mainly concern the reproduction of units, even if the gauging of measuring tools is an important part of making experiments. Besides the fact that physical theories are invariant with respect to units, a competent experimenter mainly requires a kind of knowledge which might roughly be described here with the help of simple examples: he needs to know that (and why) a stick of steel is better than a strip of rubber for measurements of length, that a quartz-clock is better than a cuckoo-clock for measurement of time or generally speaking, that a rigid rod is better than an elastic one for length measurement, a uniformly running clock better than a clock not working at all or not running uniformly for time measurement and so on for all physical parameters. The knowledge, however, of how the meter, the second, the pond, the coulomb etc. are defined is not indispensable for making experiments. In other words, the *competence of an experimentalist* is a matter of knowing which forms (like, e.g., geometric forms which allow of a definition of congruence and rigidity, and as opposed to properties of magnitude) measuring tools must assume. Therefore *theories of measurement* which do not suffer from the naturalistic fallacy have to be articulated as *systems of norms* for the production or use of measuring instruments. As such they are neither empirical nor analytical theories about the actual use physicists make of measuring instruments because these would be neither able to found nor to legitimate the claim that measurements are repeatable just because the properties of measuring instruments are reproducible. Normative theories of measurement, which control the human actions of the construction and the use of measuring tools are put forward under the name of *protophysics* (5). Such theories, however, will not be discussed here. Only one point which concerns the relation between natural science and technology needs to be mentioned.

Of course, measuring tools do have natural properties too. Such natural properties can be irrelevant, negative or positive factors for the functioning of the measuring tool. For example, the colour of the material of a clock is irrelevant, the expansion of the material of the pendulum by heat is a disturbing factor, and great hardness in the material of gears is a supporting factor in the functioning of a pendulum clock. The knowledge, however, which natural properties of a measuring tool play which role presupposes a system of norms for production and use which defines the specific *function free from disturbances*.

This remark is directed against an interpretation of the properties of measuring tools as governed by natural laws which became common in relativistic physics. On this interpretation, geometrical properties of measuring tools of length are conceived of as non-euclidian afterwards although the tools are produced in accordance with the propositions of euclidian geometry. In other words, the propositions of the relativity theory which claim to be based on measurements are estimated more highly than the technical norms which make those measurements possible. This strange, although commonly acknowledged view overlooks the fact that a single measurement can be imprecise only with relation to a particular technical purpose. This means that only where a measuring tool deviates from a norm for undisturbedness and that this deviance is technically registerable it does make sense to speak about the difference between the apparent and the 'true' property of a measuring tool. Consequently the historical improvement of measuring tools cannot be regarded as an approximation to a standard given by natural laws but only as a performance ruled by technical purposes even if the *improvement* hinges on the use of empirical scientific knowledge.

(c) Experiment

Experimenting is usually described as being guided by an interest in true empirical theories. Experimentation is called the art of finding natural laws by inquiring into nature under pure conditions. Modern textbooks for the philosophy of science give different descriptions of the experiment as a means for research into nature. Generally, however, experiments are characterized in close connection to the goal of formulating *theories*. But history shows that the interest in theories which are later called true, corroborated, successful etc. and regardless of whether they are called systems of statements or not, is a derived interest. A high respect for academic traditions on the one hand and a certain disdain for technical application on the other led historically to the belief that experiments are *nothing but* a means for gaining true knowledge of nature although experiment played an important role from the very beginning of men's use of tools and caused the development of a technology independent of any theoretical claims.

Experiments therefore have mainly to be attributed to the history of *technology*. The primary interest guiding an experimenter is a certain technical effect. This effect, or, in other words, *the function of the experimental device*, is the criterion guiding the construction or composition of this device. Of course the word 'experiment' emphasizes the fact that experiments are used for the acquisition of empirical knowledge; but even if an experiment is not involved in a search for theories, it still makes sense. To be sure, any production of technical effects causes a growth of knowledge, sometimes even as a kind of side-effect, and this growth of knowledge can be formulated linguistically. But it is crucial to see that the interest in the growth of knowledge has no consequences for the choice of the experimental device which would go beyond the measures based on the interest in the technical effect. This view of experiment can be supported by both historical and by systematic reasons.

Historically speaking, actions which can be called experiments are certainly older than any theory. This statement of course makes use of the word 'theory' in the sense of a set of sentences, and of 'experiment' in the sense of human technical actions which includes the use of tools and from which emerge certain knowledge. A simple and even a prehistoric technology then requires a *tradition*, i.e., in the sense of the Latin word *traditio*, a kind of passing on to others of the knowledge about successful or unsuccessful attempts. It is rather misleading to call this verbal passing on 'stating a theory' because a *traditio* has more the character of *instructions* than of assertions. In craftsmanship, it consists in the instructions given to the apprentice by the master. Even non-verbal abilities which can be passed on by showing and imitating belong to such a fund of knowledge. On the basis of such a knowledge, inherited from older generations or collected by himself, an experimenter can make a *plan*, again verbally or by a drawing or even with a vague hunch, to produce some new technical effect. Any attempt to realize a plan leads necessarily to success or failure. Whether this 'result', in the sense of a conscious realization of success or failure, is applied elsewhere does not have any effect on the experiment itself. The application is no criterion for any distinction between experiments as means for the construction of theories and experiments as processes of technical production.

The history of technology was not guided by theories for centuries and in some areas even today it is still not. Such a history therefore is better reconstructed from the prehistoric abilities of craftsmen who realized the technically possible as a basis for the technically desireable. In other words the *history of technology* has to be written as a *history of technical aims* on the basis of respective technical capacities and dependent on nontechnical needs.

The decisive antinaturalistic argument consists in the fact that any experiment can have a result only with respect to a criterion for success. And, for systematic reasons, this criterion cannot be the property of being an *experimentum crucis* for a theory (or being an attempt to falsify a theory). Since experiments, on the view presented here, belong to the history of technology, it has to be pointed out which role the experiments can play in *natural science*. But before discussing this question we need to ask whether any knowledge gained from experiments is properly called theory.

(d) Theory

Theories are the objects most thoroughly discussed in the philosophy of science. Here only one question shall be dealt with in order to criticize the naturalistic understanding of theories: *What are theories formulated for?*

According to the view developed above, success or lack of success of an experiment causes a growth of knowledge for the experimenter. Then the problem of preservation and of passing on this knowledge arises and under certain circumstances an explicit formulation of what is held to be the result of the experiment may be needed. This all sounds rather obvious but is commonly disregarded as far as the consequences of the purpose of theories, namely to make tradition possible, is concerned.

Regardless of the many possibilities for further differentiation of theories with the help of logic, theories of definition, syntax, semantics, model theory, etc, theories also have to be considered as means for making knowledge communicable, that is, whether they allow the acquisition of knowledge just by learning what teachers teach or colleagues tell. The 'truth' of a theory (or the corresponding concept according to the different views) which has become the central topic for philosophical discussion, has to be subsumed under the purpose of making knowledge *teachable*. In other words *theories* have to be judged according to the criteria of how *suited* they are *for communication* and *how instrumental* they are *for technical effects*. And yet theories based on experiments need not fulfil the requirement of being 'true pictures of nature' or 'natural laws'.

One important methodological consequence of this view is that a lot of philosophical debate can be dispensed with – for instance the quarrel between inductivist and deductivist views, or the discussion whether theoretical terms are definable, completely or uncompletely interpretable etc. An especially important consequence is that the question about the appropriate concept of truth for scientific theories can be answered simply. There are also positive methodological implications like, for example, the postulate that any scientific terminology has to be built up explicitly and step by step in order to be teachable. Finally it should be noted here that *theories of physics as a natural science* in the stronger sense, i.e., as a science that does not make experiments but observations of *natural* phenomena, may have a different status than theories of the domain of experiments which serve technology because of their systematic presuppositions pointed out above.

(e) Technology

The main conclusion of the passages on measurement, experiment and theory was that *major parts of physics* which are attributed to natural science are shown by a closer analysis to be *technology*. A look at the presuppositions of the art of making experiments and at the formulation of theories in connection with experiments suggests the classification of physical knowledge as a knowledge how to effect things technically. A survey of the different branches of modern technology allows us to specify tasks which belong to the realization of technical knowledge in accordance with traditional specifications of branches of physics.

Two aspects will be discussed here to illustrate the previous sketch of an non-naturalistic philosophy of physics. There are two particular fields of technology which have to be looked at from a methodological point of view, namely the *technology of measurement* and the *technology of observation*.

'Technology of measurement' means the highly developed instrumental support of measurement in modern experimental science. Anyone who is interested can inform himself in a modern physical laboratory – if he is prepared to admit that he already requires a lot of physical knowledge in order to be able to understand how physicists gain physical knowledge. This seems to be a logical circle; in fact it is not. Methodologically speaking, the technology of measurement *aims at more refined ways to reach the goals of measurement*. The goals of measurement in the most general sense is to state ratios of magnitudes ('parameters') in a controllable way – for instance, in order to reproduce experimental conditions. The controllability consists mainly in the reproducibility of the relevant properties of measuring tools. Technology of measurement therefore has to be seen as a knowledge *how better to cause* the reproducibility of relevant properties of measuring tools. Yet the technical improvement presupposes logically a criterion for improvement. In other words, the uncontroversial fact that modern art of experimenting is based on a highly sophisticated technology of measurement does not dispense from the need of a nonempirical criterion for what an undisturbed function of a measuring tool is.

Consequently, the historical development of the technology of measurement which naturalistically is interpreted as an approximation of the properties of measuring tools to natural laws can more appropriately be understood in the following way: a feed-back of a technology, itself grounded in the art of measurement, to the practical construction, improvement and use of measuring tools allows *a successive approximation of real properties* of the instruments *to the ideally postulated ones*.

(The postulates are argued for in general technology and in physics by appeal to the goals of measurement).

The second aspect to be discussed is the 'technology of observation', that is, the technical support for the observation of phenomena 'given' by nature or by man. Examples stretch from glasses to microscopes or telescopes, but include all observational techniques which cover information handling by systems which, physically speaking, belong to a different complex of phenomena, for instance the electronic amplification of acoustic phenomena. What was often wrongly said of measurement holds for observation: the technology of observation supports our sense-perception by amplifying the sensibility or the range of frequences, etc. Of course, observations are not an end in themselves. It depends on the goal of a particular observation whether a particular technical device really supports this observation or not.

The specific purpose of observation is to inquire into nature in the strict sense of that what is not made by man – planets and stars, the weather, the northern lights and the gulf stream, earth quakes and sun rays, and so on. (There are also 'organic' phenomena in which a distinction between physical and, say, biological observations is hard to make.) Methodologically, it is important to see that *observations* of natural phenomena made with apparatuses presuppose a *technology* of observation and that this technology is *based on an experimental science* of artificial and therefore known phenomena. It is therefore misleading to assert that physics as a natural science is based on observations and to neglect the distinction between every-day and scientific observations, i.e., observations by means of tools with reproducible properties and to propose that technology and its branch of observational technology is a side-effect of physics. *Physics as a natural science is possible only on the basis of physics as a technology.* Technology is a presupposition of any scientific knowledge which is claimed to be based on *repeatable* observations. Repeatability can only refer to the function of the observational tools. Whether the natural phenomena are repeated is, by their very definition, not subject to human intervention.

It now remains to show in which sense physics, which on the one hand has to do with technology and on the other hand with natural phenomena in branches like astronomy or meteorology, deals also with 'nature', that is, describes states of nature 'natural laws'

(f) Nature

Up to now the word nature meant the domain of phenomena not caused or produced by men. It was mentioned at the outset that the change in the meaning of 'nature' in connection with classical mechanics gave rise to the misuse of the word 'nature' for artificial objects. In a liberal reading of the Latin or Greek roots of the words 'nature' and 'physics' respectively one can disregard the organic aspect and simply call all phenomena *natural* which are not a result of human actions. This proposal is not terminologically sharp because it leaves unclear whether non-intended side effects of intentional actions (like pollution) are natural or artificial. One could therefore extent the terminology by calling those side effects artificial *which could be an end* for a human action (even if an unreasonable end). Another problem has to be left open here. There is human behavior like breathing which is not an action in the sense that it is not intended, not oriented towards certain ends, and not judgeable with respect to success. Such behavior is natural. But there are borderline-cases where it is difficult to say whether a certain event is a human action or a human behavior, i.e., whether the action is intended or just occurs. But this terminological question which is still open does not influence the answer to the question whether physics is a natural science or not.

In the light of the use of 'natural' proposed just now, physics is a *natural science* just then when it deals with natural phenomena. These branches of physics are based on observations, but not on experiments. The other branches belong to technology. It should not be overlooked however that even the natural-scientific parts of physics are not ends in themselves (since man were be born curious and heirs a satisfaction when he discovers the secrets of nature). Astronomy is a supplementary natural science for navigators and calendar makers, meteorology a supplementary natural science for farmers, sailors, pilots, travelling agencies, and so on. And knowledge about nature can take the role of an emancipating force against myths about nature. Both purposes of natural science in the strict sense, the practical and the emancipatory one, can be pursued only by means of a natural science based on scientific observations in the sense of repeatable observations. Or, to put the same point more sharply: no practical or emancipatory use of knowledge about nature without scientific observation, and no scientific observation without technical knowledge about the instruments for observation.

This part, the branches of physics being natural sciences in a strict sense, forms the smaller problem. The common opinion emphasizes more *the lawlike character of physical statements* outside the field of *technical* knowledge, or, for short, that technology is possible because of natural laws. There, one argues, it is just the success of technology which proves the existence of natural laws.

This naturalistic position can now be reconstructed critically in the following way. In the previous discussion of the technological parts of physics, we touched upon the gain of technical knowledge, about the forming of traditions of research, and about technical ends or purposes. The fact however that not everything that can be wished for or aimed at by human fantasy in the field of technology can technically be realized has not yet been taken into account. Some technical enterprises are successful, others are not. The more one understands about technology the more one knows about the limits of technology, too. This knowledge can be formulated in a metaphorical way as a meta-statement about the statements (axioms, theorems) of physical theories, that the (technical!) success of those statements proves the lawfulness of nature. Although it might be misleading to speak about 'lawfulness' because the 'law' suggests a lawgiver, it has some plausibility with respect to every-day language to say that it is nature which limits technical means and in precisely this sense, that *nature is accountable for the success or failure of particular technologies.*

The alternative to naturalism then is this. Theories of physics, even of a technological character, may fail – metaphorically speaking – because of nature. But, on the other hand, this failure necessarily presupposes explicit criteria for technical success. Therefore *knowledge about nature* in the technological parts of physics is *not*, as opposed to the naturalistic view, *independent of history*. Physics can be understood only in the contexts of cultures because it is based on instrumental experience.

(g) *Policy*

In some methodological arguments against naturalism, some consequences are briefly mentioned which go beyond the field of methodology. As it cannot be the purpose of this paper to develop a program for a policy of science or research, only those political consequences shall be enumerated which arise from the criticism above.

If theories are not held any longer to be true pictures of the natural world, if theories are construed as serving the task of making knowledge communicable and traditions possible, the criterion of teachability must gain more importance than in contemporary physics education at high schools and universities. In general, the so-called exact sciences, which follow the ideal of axiomatic theories in a formalistic understanding of sciences, suffer from a difficulty which analytic philosophy of language has properly described as the problem of theoretical terms and their definitional deficiences. In the teaching of these sciences *no complete, explicit, and step by step way of introducing a physical terminology exists.* The student has to pick up terminology like a jargon; when he receives his degree he realizes that he is 'in' but he cannot give definitions for the key concepts of his theories. Of course scientists invented marvellous excuses for this way of speaking and teaching the saddest of which is to emphasize the function of this jargon in discriminating between gifted and less gifted students. But even good teachers which do not follow this method have to face the fact that the pedagogically required definitions for the terms of physical theories are just unknown. This of course will not change so long as physicists espouse the formalistic program of 'interpreting' formal axiomatic theories (as if their authors would already have died) instead of making theories complete by giving definitions. In any case, the argument that these definitions are impossible because of the complexity of the object ('nature') of physical theories can be proved wrong. A non-naturalistic or a culturalistic understanding of physics would lead at least to the program of explicit terminologies.

A culturalistic orientation of physics has even more important consequences for *the policy of research planning* than for science teaching. It is customary to distinguish, mainly in political debates, between financial support for applied research on the one and fundamental research on the other. This distinction supposes that applied physical research is guided by external goals whereas fundamental research is neutral with respect to application. There are different defensive strategies to fight the question for the purposes of fundamental research. To mention just a few: the question would be senseless by definition and therefore dull. Fundamental research of to-day is applied research of tomorrow (which implies that all knowledge is finally applicable in a good way). Fundamental research would produce knowledge in reserve, and history shows that such knowledge is ever useful. Fundamental research finally would contribute to human knowledge which is a value in itself, and would therefore contribute to culture, civilization, etc. These defensive arguments *de facto* have the function of opening room for fundamental research without assuming the obligation of legitimating such research.

These possibilities, however (the restriction of which is not my purpose to argue for here) do not exclude judging the activities of scientists in the field of fundamental research according to criteria of the division of labour in a highly developed society. Such judgements which, of course, should finally lead to distinguishing suitable projects from less suitable ones can be seriously discussed only if the assumption that it is generally good to inquire into nature is no longer accepted as an argument.

A further political consequence of a culturalistic understanding of physics is to diminish the *schism between natural and cultural sciences* at the universities. This schism leads to mutual exclusion of certain contents and implies that so far as their education is concerned natural scientists suffer from political incompetence, whereas social scientists suffer from mathematical and technical incompetence. If however physicists do not any longer see the solidity of their science in the neutrality of the object of their science with respect to history but acknowledge the dependency of their *de facto* technological research on general historical and political conditions, then the eccentric picture of a scientist looking for truth about nature like a romantic visionary could be replaced with an understanding which sees the role of responsible physicists in contributing to solutions of global social problems.

According to the non-naturalistic view of physics, scientists cannot argue that they just discover natural laws and that it is only a matter of political decisions about *applications* which determine whether this discovery is useful or destructive. They have to be involved as the scientifically competent individuals in means-and-ends discussions about research programs. Of course, it can happen that means chosen for acknowledged ends have noxious side effects. But the problem today lies more in the fact that means-and-ends discussions do not have a place at all in the problem of choosing between the advantages of ends and the disadvantages of side-effects.

Although it was the purpose of this paper to sketch a non-naturalistic philosophy of physics in the field of methodology, it may have become apparent that a criticism of naturalism is not merely a quarrel about trifles but has important practical consequences.

Notes

1. The program to found geometry by a system of norms prescribing operations to bring about ('geometric') forms of bodies like planes, straight edges or perpendicular pairs of planes, has been worked out in P. Janich, 'Zur Protophysik des Raumes', in G. Böhme (ed.), *Protophysik*, Frankfurt: Suhrkamp, 1976, pp. 83–130. The program itself goes back to P. Lorenzen and owes central ideas to H. Dingler, Cf. P. Lorenzen, 'Das Begründungsproblem der Geometrie als Wissenschaft der räumlichen Ordnung', in P. Lorenzen, *Methodisches Denken*, Frankfurt: Suhrkamp, 1968,[2] 1974, pp. 120–141. K. Lorenz and J. Mittlestraß (eds.), *Hugo Dingler, Die Ergreifung des Wirklichen*, Frankfurt: Suhrkamp, 1970.
2. The protophysical program has first been worked out in the field of time-measurement. Cf. P. Janich, *Die Protophysik der Zeit*, Mannheim: Bibliographisches Institut, 1969. A revised and enlarged edition will appear in 1978 ('Protophysics of Time', in *Boston Studies in the Philosophy of Science* **XXX**, Dordrecht: Reidel).
3. This program is first formulated in P. Janich, *Zweck und Methode der Physik in philosophischer Sicht*, Konstanz: Universitatsverlag, 1973.
4. A detailed argumentation proving this statement is given in P. Janich, Eindeutigkeit, Konsistenz und methodische Ordnung: normative versus deskriptive Wissenschaftstheorie zur Physik', in F. Kambartel and J. Mittlestraß (eds.), *Zum normativen Fundament der Wissenschaft*, Frankfurt: Athenäum, 1973, pp. 131–158. An explicit critique of empiricist theories of metrization is developed in 'Protophysics of Time' (cf. Note 2).
5. The term 'protophysics' (although occurring already in the 16th century in different connections and senses) was first proposed by P. Lorenzen in the sense meant above. Cf. P. Lorenzen, 'Wie ist die Objektivität der Physik möglich?', in P. Lorenzen, *Methodisches Denken*, pp. 142–151.

GOAL DIRECTION OF SCIENTIFIC RESEARCH

RON JOHNSTON
Liberal Studies in Science, Manchester University

and

TOM JAGTENBERG
School of Sociology, University of Wollongong

1. Introduction

That science, both as an institution and a body of knowledge, is neither invariant nor homogeneous is by now widely accepted. Thus Ravetz (1) has argued that three distinct varieties of science have been historically perceived and have been associated with particular interest groups:

(i) 'Pure, academic science' which was based on an ideology derived from nineteenth-century German universities:

Here science is totally inward looking, its only offerings to the outside world are general contributions to knowledge and culture, unpredictable technological applications, and the example of its endeavour (2).

(ii) 'Ideologically engaged science', which is considered a bearer of truth and reason, standing against dogma, superstition and oppression, and a weapon in the struggle against a variety of material and spiritual ills:

By its very nature science could not produce either error or evil, and so it had a privileged position among all sorts of ideologically engaged activities (3).

(iii) 'Useful Science' where 'the results and methods of science are applied directly to technical and practical problems; and those external tasks provide stimuli, goals and partial justification for scientific work' (4).

There is considerable evidence that it is the last of these which is most appropriate as a description of contemporary 'science'. Research has become concentrated in industrial and government laboratories (5), where the demands for profit and growth and accountability, respectively, require that the research be directed, at least in the long term and more

29

Krohn/Layton/Weingart (eds.), The Dynamics of Science and Technology.
Sociology of the Sciences, Volume II, 1978. 29–58. All Rights Reserved.
Copyright © 1978 by D. Reidel Publishing Company, Dordrecht, Holland.

often in the short term, to practical ends. Furthermore, it is directed primarily towards the general objectives of economic development and national security (6).

This orientation towards 'useful science' has had a considerable effect on the status and self-images of those engaged in scientific research, whether they call themselves scientists, research workers or technicians. The growth of unionisation among scientific workers (7), reflecting an increased awareness of their loss of status can be seen as a response to the directed nature of modern research. Furthermore, this direction is not restricted to the environment of government or industrial laboratories, as shown by the organisational constraints operating in 'big science' institutions such as CERN (8).

In spite of the evidence that research has become a much more highly controlled and directed activity, academic scientists and in particular the elite who represent the scientific community in negotiation with the paymasters and political interests, continue to project the image of science as autonomous and apolitical, concerned solely with the objective pursuit of truth, i.e., as 'pure academic science'. A clear illustration of the solidarity with which this position is held is provided by the response of British scientists to the Rothschild reorganisation (9). Such a view can be considered a self-serving ideology, protecting the interests of scientists from external control (10):

> The theory that scientists follow only the internal rules of science would seem to reinforce their efforts to prevent the subordination of their work to standards extrinsic to science and to protect themselves from external political influence (11).

We would wish to argue that the continuation of this misleading view of science represents a dangerous mystification which can serve only to obscure the role of science in modern society, confuse the scientist and science administrator, and alienate the public.

The extent to which this protective self-image has influenced the study of science and its institutions has been shown by the pervasiveness of the analytical dichotomy between 'internalist' and 'externalist' explanations which reinforces the notion of science as a social system separated from, but occasionally influenced by, other social forces (12). It is also evident in the concentration of studies on precisely that small sector which most nearly approximates this model – that of academic, university-based, 'pure' research (13).

This disjunction between 'pure' and 'useful' models of science is reflected in the conceptual distinction between 'pure' and 'applied' science (14) which has provided the basis for most attempts to relate scientific knowledge to socio-economic objectives. In this model, certain 'pure' kinds of research involve the objective pursuit of truth, and should be conducted in entirely autonomous fashion, i.e., according to the regulatives of science only. Other 'applied' research is directed to the achievement of specific and practical objectives and may therefore be expected to be administered and held accountable, in more or less the same way as any other production or social function (15).

The establishment and subsequent growth of science policy, with its emphasis on mechanisms for evaluating claims for research support, and the form of institution most suitable to administer research funds, reflected a concern to more efficiently direct science to desired ends (16). However, once again the self-image of science appears to have been accepted:

The underlying model for science policy organisation is based on the transaction concept drawn from political science; its aim is the establishment of effective institutionalised transaction processes between an independent science institution and society, via society's representatives in government. In this model there is no mechanism permitting society's interest to operate on the scientific institution, and analysis, planning and one might add, responsibility, is limited to the areas of *application* of scientific knowledge (17).

Blume (18) has argued for the establishment of a political sociology of science directed towards the explanation of the contemporary politically directed, occupationally differentiated and institutionally disparate form of science, but as yet there has been little direct response to this challenge. There is, of course, a well established Marxist tradition of relating science to social needs which can be traced back to Hessen (19) and has included within its ranks such notables as J. D. Bernal (20) and Joseph Needham (21), but its major contemporary expression is in detailed historical studies of the way in which particular scientific theories reflect the socio-economic and cultural context (22).

One promising development has been the work of the project group 'Alternativen in der Wissenschaft' at the Max-Planck-Institut, Starnberg, and Weingart at Bielefeld, who have attempted to develop the concept of 'finalisation' to establish theoretically the conditions for effective direction of scientific knowledge towards politically determined goals (23). This approach has been developed over the past five years and in the most recent paper (24) takes the form of a sophisticated and fruitful model for the formation and transformation of research objectives, and the limitations placed on the achievement of such objectives by the state of development of relevant knowledge fields.

However, although the supporting case studies consider a range of specialties (25), we wish to argue that 'finalisation' represents only one extreme, and rather unusual type of externally directed science, taking as its implicit model the U.S. crash programmes to land a man on the moon (the Apollo project) and to find a cure for cancer (the 'cancer moonshot'). There may be a very wide spectrum of goal-orientation within science, ranging from the directly politically motivated programme to a much more general and pervasive shaping of cognitive goals in accord with social or economic needs.

Thus there is a need to develop a more general analysis of the establishment and operation of goals both within the scientific community and between it and its paymasters. Such an analysis should take account of the fact that scientists are themselves involved in the process of goal formation, and that it is determined, at least in part, by the state of knowledge perceived or claimed to be relevant. Furthermore goals, at least at an operational level, may be in a state of continual transition and redefinition. Such an analysis of the function of goals in science is seen as part of a programme to develop a flexible theory of mediation between socio-economic and cognitive aspects of the social reality which constitutes and conditions science. Previous papers have explored the need for a contextual model of scientific development (26) and the roles of occupational and cognitive differentiation in shaping the knowledge structure (27).

2. The Common-Sense Notion of Goals for Scientific Research

All scientific work can be regarded as goal-directed, at least in the sense of Habermas (28) who argues that all work is 'purposive rational action', in that it is either instrumental action, or rational choice, or their conjunction. Thus,

science is oriented towards goals which may be directly or indirectly perceived, which are expressed at varying levels of generality, which are developed and transformed by social processes and which are dynamically linked to scientific practice and the state of knowledge in such a way that the goals and science are only analytically separable (29).

However, the fact that *all* action is directed to goals would appear to have led students of science to treat goals as a common-sense phenomenon. As a consequence the 'low-level' goal of problem-solution has been the primary focus of investigation, and other levels or kinds of goals which may shape the research process have been neglected. More-oever, the existence of 'problems' has been taken for granted; in the positivist model of science, problems are posed by the external reality. It is only with the development of the post-Kuhnian sociology of scientific knowledge that the perception of a scientific problem has been considered in need of sociological explanation. Furthermore, the acceptance of goals as a common-sense phenomenon has entailed a common-sense relation-ship between goals and the means used in the attempts to realise them. Consequently the relationship between means and goals (or ends) has not been an issue in science studies. The dominant approach has been a functionalist analysis of the evolutionary movement of means towards 'immaculately conceived' goals.

The analysis of goals by sociologists, at least in the sense used in this paper, has in general been quite limited, with the marked exception of organisation theorists. Within this literature, there are a number of distinct traditions, including of relevance to this paper, studies of the range of goals pursued by industrial firms, the evolution of goals within organi-sations, and means of changing goals in normative institutions (30). We will draw on some of these concepts in subsequent arguments.

As already noted references to goals within science are in general sparse, or made, in passing, as part of a more general anslysis. Thus Merton (31) sees the goal of science as 'the rational pursuit of truth'. For Sklair (32), 'the charter or purpose of science' is of three types *viz.*, the quest for knowledge for its own sake, for alleviating human suffering and satisfying the needs of mankind, or to provide an economically rewarding career. Richter suggests:

The goal of science, as commonly recognised today, involves the acquisition of systematic, generalised knowledge concerning the natural world; knowledge which helps man to understand nature, to predict natural events and to control natural forces (33).

Ravetz has made a considerable effort to clarify the notion of goal by developing a hierarchy of 'final causes' that determines the goals of the research task. He distinguishes between goal, function and purpose:

the task itself has a goal, which is conditioned more or less strictly by the function which will be performed by the result of the accomplished task; and this in turn is governed by the ultimate human purposes which are expected to be served by the performance of that function (34).

The goal is the solution of the research problem.

The most detailed analysis of goal-direction in science is offered by Hagstrom (35) but in general the concept is used in a common-sense way to explain structural change, the role of fashion and the processes of disciplinary differentiation and social control. Thus, although Hagstrom observes that:

Segmentation begins with cultural change, the appearance of new goals in the scientific community. Of course, new goals do not spontaneously appear: scientists actively seek them (36),

he provides few indications of the origin of the goals and continues by examining the way scientists respond to cultural change once it has occurred. The uncritical use of the concept is highlighted by the conflation of 'goals' with 'problems'. Thus, the above quotation continues,

Those who discover important problems upon which few others are engaged are less likely to be anticipated and more likely to be rewarded with recognition (37).

Where do these goals come from, and how are they formed? Hagstrom, adhering to the internal/external demarcation, distinguishes between goals arising inside and outside science:

When the relative importance of goals is easily ascertained by generally accepted criteria, or when the goals are given by non-scientists, there will be little play of fashion. In many of the applied sciences, where the goals arise outside of science and the criteria of success are usually given by non-scientists, scientific fashion is perhaps least important. In the empirical sciences, especially those with a more or less rigorous theoretical framework, the goals arise within science, but in many respects they appear to be 'given' in the confrontation of theories by 'nature' (38).

The change of goals directed from within science is due primarily to the actions of leaders:

The orderly succession of goals in a discipline is the sum of individual responses to a situation being changed by discoveries. Changes in the goals of individuals are facilitated by the tendency of scientists both to seek social validation of their goals and to follow the lead of out-standing men . . . . The ease with which physicists can change the goals of their discipline is linked with the structure of leadership in the discipline. While the ease of determining the really important problems makes it easier to spot leaders, the existence of leaders facilitates the orderly succession of goals (39).

Hagstrom makes other distinctions between types of goals, but not in any systematic fashion. Thus goals may be 'short term', i.e., specific problems being researched (40), 'traditional' disciplinary goals, as for example the purely biological goal of understanding life as a function of the cell (41), 'applied goals' such as the pursuits of industrial and government laboratories (42) and motivational goals such as incentives, particularly recognition (43). Implicitly, all the objects of competition between individuals and organisations are treated as goals. For example, position, promotion, research facilities and graduate students are scarce resources to be competed for, i.e., goals to be achieved (44).

The recognition of different levels and types of goals represents a considerable advance, but by failing to distinguish between goals and problems, and even more by linking this analysis with Merton's (45) five ideal types of scientific performer Hagstrom is committed to a static analysis which cannot comprehend interpenetrating goals and means. More fundamentally the analysis of goals, in terms of purposes to be achieved, is hindered by the adoption of a functionalist stance, whereby goals and changes in them can be interpreted only in terms of maintenance of the institution and correspondence with the means for achieving them.

3. An Exposition of Goal-Direction in Scientific Research

Before proceeding further it is important to clarify the meaning to be applied to the concept of goal. It has been noted previously that this concept has not been rigorously developed, in part because almost any sequence of behaviour can be divided into segments of varying 'size', each of which can be said to be goal-directed:

Since there is only a relative distinction between means and ends, and since, therefore, any end or goal can be seen as a means to another goal, one is free to enter the hierarchy of means and ends at any point (46).

While precise semantic definition may be difficult, in this paper we will attempt to restrict the usage of the term goal to an over-arching objective or end which serves in part to direct one or more fields of research and of which individual, or groups of, scientists need not be explicitly aware.

It is important to recognise that we are not here considering the goal of 'science'. Along with Richter (47) we reject the notion that the concept of goal can be fruitfully applied to the whole enterprise of science. Recent developments within the sociology of scientific knowledge (48) emphasise the distinctive cognitive and social structure of different *sciences*. However, whether science is examined from a cognitive perspective, be it discipline, specialty or research area, or from an occupational standpoint, in industry, government or university, we would argue that the concept of goal can provide a very useful insight into the ways in which the modern sciences reflect and can be directed to the needs of industralised society.

Furthermore the distinction that Ravetz (49) makes between goal and purpose can be accounted for, in that each scientist may structure his or her research in order to achieve particular individual purposes, but goals form an important, and so far unconsidered, element of the cognitive, and hence social, structure, which shape the possibilities of achieving particular purposes.

In this paper, a clear distinction is made between goals and research tasks or problems. It can be argued that much of the sociology of science has been directed to an examination of the cognitive and social means of achieving the low-level objective of problem solution. Considerable progress has been made in explaining how cognitive and social structures, in part determined by the elite members of a research community (49a), shape the range of appropriate research tasks open to the 'autonomous' academic scientist. The way in which the apprentice, from student to post-doctoral fellow, is presented with a research 'package' which closely defines an appropriate set of research problems has been usefully explored by Whitley (50) and others. Less attention has been applied to the work of scientists in industrial and government environments, though here it is clear that the research tasks are equally prescribed, though in a more overtly hierarchical or bureaucratic manner and with a more immediate orientation to the objectives of the organisation. While studies of the research task are of undoubted value, we are concerned here with the extent to which high level goals form part of the structure determining these research tasks.

Particular goals may vary greatly in type, level of application and origin. The development of Kuhn's models suggest that cognitive structures operate at different levels and thus it appears reasonable to infer that goals, which to a large extent will express themselves through the cognitive structures, may also operate at different levels. Using Whitley's categorisation, at the highest level goals form part of the 'metaphysical' component of scientific knowledge – 'the overall system of values and beliefs which serves to justify and integrate the scientific activity with other systems of production . . . and provides a general world view' (51). Such high-level goals need not be consciously associated with all phases of scientific work; they may form part of a tacit background knowledge inter-nalised through socialisation processes of which the scientist or administrator may be unaware. Nevertheless they may, in a highly mediated form, provide a powerful directing influence on scientific research. To illustrate, Mullins has identified the high level goal of the phage group as determining 'the secret of life' (52). On the basis of a detailed examination of solar energy research Jagtenberg has argued that the consensual high level goal has been "making the sun work for mankind by extracting large amounts of useful energy from solar radiation" (53).

Goals may also be formulated directly at the next level of cognitive structure, i.e., specialty concerns, described as: "the general problems or purposes of conducting the activity seen in terms of a particular definition

of reality which may incorporate a number of evaluative frameworks" (54). At this level goals may take a rather more concrete, and
explicit form as, continuing the phage group example, the determination
of "the mechanisms by which genetic information is transferred" (55).
Other examples include plasma physics, where the goal is one of understanding the properties of plasma sufficiently to allow continued
controlled fusion with a positive energy balance, or biotechnology, where
the goal is the artificial mutation of microorganisms suitable for the
manufacture of industrial products (56). Cognitive structures at lower
levels, namely explanatory models, techniques and research practices,
may contain expressions of the higher level goals but the latter are unlikely
to operate directly at these levels.

Goals may also be of rather different types. Thus some goals may take
the form of very highly mediated expressions of socio-economic and
cultural context such as operate in physics and biology (57). At the other
extreme goals may be much more direct expressions of non-scientific
interest groups such as sectors of government, industry, or the public, as
seen in the development and direction of fields such as computer science,
geology, tribology, toxicology and environmental science. We emphati-
cally reject the adherence to the internal/external dichotomy which leads
van den Daele *et al.* (58) to restrict the concept of goal to an objective
developed outside the scientific community. In accord with the view of
scientific knowledge as the highly mediated expression of socio-economic
as well as ontological constraints, the analysis of goal-direction should be
sufficiently general to apply to all sciences.

The major argument of this paper is that all modern science can be
understood as a goal-directed activity. That is, scientists and those concerned with directing and influencing research operate according to goals
which are

(1) Established as the result of social and political processes which
involve dynamic interaction between interest groups both involving and
excluding direct scientific interests and which may be directly or indirectly
perceived by scientists.

(2) Mediated by scientific, socio-economic and political considerations
and expressed at varying levels of generality; (59) these mediated versions
may be expressed within 'official' statements of research programmes or

they may be deeply embedded in the cognitive structure of the relevant sciences.

(3) Dynamically linked to an evolving body or bodies of scientific knowledge in such a way that goals and science are only analytically separable; both cognitive and social aspects of research are directed and constrained by orientation to goals which are posited and potentially continually redefinable in terms of changing theory, techniques and socio-economic conditions.

In summary, three characteristic features of the goal-direction of scientific research can be identified: goal establishment, goal mediation and goal evolution. It is important to recognise that these are only analytically distinct; it would be very unusual for them to follow each other as three distinct phases. In the subsequent sections we will examine these features and illustrate them with examples drawn from a number of case studies in order to assess the utility of the concept of goal-direction in science and to develop a model which provides the basis for a clearer and more practical understanding for both scientists and science administrators of the ways in which modern science is, and can be, directed.

4 Goal Establishment

Van den Daele *et al.* have pointed out with considerable clarity that the establishment of goals – or as they describe it, "the transformation of political into scientific goals" (60) – can involve a varying degree of determination by political and scientific authority. The establishment of goals for scientific research programmes have been thoroughly documented for such classic cases as the Manhattan Project (61), the Apollo Mission (62) and the U.S. Cancer Programme (63). However, below are presented studies of goal establishment in less spectacular and possibly more typical cases of goal-directed scientific research.

Tribology (64)

Concern with the problems of friction and the means of eradicating it is not new. The American Society of Lubrication Engineers was formed in

1944, and by 1954 the subject of lubrication engineering was being taught to undergraduates at Imperial College, London (65). However, Dowson (66) has argued that the industrial developments of the post-war period posed a series of qualitatively new questions for mechanical engineers. High speed rotating machinery, more severely loaded reciprocating machinery and hostile working environments were increasingly causing breakdown and failure in production hardware. In 1962, a special committee was set up by the U.K. Department of Scientific and Industrial Research (DSIR), under the chairmanship of Mr. G. B. R. Fielden to investigate 'the problem' of design within mechanical engineering. At the time, the engineering industries accounted for 35% of manufacturing industry, nearly half of the United Kingdom's total exports and was the main supplier of domestic plant and machinery. The Fielden Report (67) argued that the industry was not modernising sufficiently quickly, and that it was particularly lagging in the fastest growing, technically more advanced sectors. Vig has argued that the

Fielden Committee's frank report and DSIR's follow-up campaign received much public attention and contributed to a growing consensus on need for further reform of education, management, and State support programs (68).

By the early 1960s a Lubrication and Wear Group had been formed within the Institution of Mechanical Engineers, and it was within this group that some of the aforementioned problems were highlighted. Moreover it was at this time that science and technology became important issues in British politics (69). Two of the more strategic battle-cries of the successful Labour Party were the welding of science and socialism and the modernisation of Britain's industrial structure. The latter argument was particularly resonant with the state of affairs which mechanical engineers were facing in the fields of friction, lubrication and wear and the development and use of such knowledge in design procedures.

The crucial factor in the emergence of tribology as a distinct field of research was the generous support it received from the government. Mr. Peter Jost, a prominent member within the Lubrication and Wear Group of the Institute of Mechanical Engineers, was acquainted with, or at least 'had the ear of' Lord Bowden who was then Minister of State at the Department of Education and Science. Given the Labour government's manifesto, Bowden was probably sufficiently swayed by Jost's arguments that a lot of money could be saved in this crucial industrial sector if appropriate steps were taken and that this would, simultaneously, be a concrete realisation of one of the major promises in that manifesto. At any rate, whatever happened in this respect, Bowden *did* ask Jost to establish a committee of experts to examine the position of British lubrication education and research and to opine on the needs of industry in this field. A working group was established in late 1964 and eventually published the Jost Report (70) in February 1966.

This report justified the importance of its subject area in two ways which were obviously deeply resonant with Labour's electoral pledges:

(1) Potential savings were estimated at £515m per annum;

(2) Longer term economic, industrial and commercial benefits would accrue if the rate of technological progress was improved by recognition of this important problem. The barriers to this progress could be removed and the reputation of British engineering goods would be enhanced.

To effect these goals, the working group argued that more research and education, and the development of a general awareness of the important potential of the subject throughout industry was required. They also argued that these benefits had not been realised because of the subjects multidisciplinary nature and 'oil-can' associations. In an attempt to remove the latter fetter, the working group thought it necessary to devise a new name. So, following consultations with the editors of the Supplement of the *Oxford English Dictionary*, 'tribology' – from the Greek 'tribos', meaning 'rubbing' – was decided upon and defined as: 'the science and technology of interacting surfaces in relative motion and of the practices related thereto'.

In February 1966, responsibility for the mechanical engineering industries was assumed by the Ministry of Technology (71). This explains why it was Mr. Wedgewood Benn, the then Minister of Technology, who announced to the House of Commons, on 11 August 1966, the formation of a Committee on Tribology under the chairmanship of Mr. Peter Jost. Essentially, it was set up to advise the Ministry as to how the recommendations of The Jost Report could be most effectively implemented. The major results were the establishment of an Education and Training Committee, three Centres of Tribology research, and a Research and

Liaison Committee which, among other functions, assessed research proposals for the Science Research Council.

Thus by a detailed political process involving engineers, industrialists and civil servants the high level research goal of understanding interacting surfaces in relative motion was established. While the goal of 'understanding', it may be argued, is no different from that of science in general, it should be noted that the conditions and justifications for the programme focus the desired understanding within a particular set of constraints which can be best described as goal-oriented. A further point, and this marks a considerable distinction from the model of van den Daele *et al.* (72), is that the goal is not directed to or emergent from one 'finalised' science. Rather is has been adopted by specialists whose interests are or have become directed within the area of tribology from such a wide range of more traditional specialties such as crystallography, metallurgy, fluid mechanics, and various branches of chemistry and physics. Thus, the high-level, 'metaphysical' goal was transformed into a more immediate 'specialty concern' goal by a process of mediation for each of the relevant specialities – a process we will examine in more detail in the next section. But first, another example of goal establishment.

Environmental Goals (73)

In the early 1960s, the first rumblings of discontent over the despoliation of the environment in the industrialised countries were beginning to make themselves heard. The environment movement originated in the U.S.A., characterised by the appearance of such books as Rachel Carson's *Silent Spring* (74) and Barry Commoner's *Science and Survival* (75). Carson's book, in particular, caught the imagination of the public with an emotive discussion of the damage inflicted upon wildlife by the organochlorine insecticides. Subsequently, the destruction of the environment became an increasingly popular theme for discussion, and in mid-1969 the 'environmental crisis' arrived. This arrival was accompanied by a publication explosion and the flood of books and journal and newspaper articles focussed upon the 'crisis' has still not completely abated. Similarly, there was a sudden upsurge in the formation of environmental action groups in America and Europe.

A number of key issues dominated the discussions on the environment, particularly the population explosion, the depletion of natural resources (mineral and energy), and the ever-increasing discharges of liquid, solid and gaseous pollutants into the environment. Less tangible worries included the increase in noise, the threat of nuclear war and fears for the general debasement of the quality of life. 'Ecology' quickly became a new watchword, despite that fact that it had originated as a biological concept in the latter half of the nineteenth century; it was seen by many to hold the key to the solution to the environmental crisis. The most important element of ecology lay in its emphasis upon the total interrelatedness and interdependence of all this planet's living matter, including man (76). Man is an integral part of the global ecosystem; his disregard of this fact was the cause of his environmental problems.

The environmental movement was by no means united. A large, predominantly youth-based faction was committed to the environmental cause because of its potential radical implications. This group's enthusiasm was probably responsible for the launching of the entire movement. In the prevalent counter-cultural spirit of the late 1960s, when the opposition to technocratic society was at its zenith, the environmental cause was another direction through which to channel anti-system energies and sentiments; the 'crisis' was seen as a damning indictment of the *status quo* and technological rationality.

The conservative version of the environmental campaign depicted the problems as technical rather than political. It has been suggested that this conservation ethos, predominantly middle-class, was based chiefly on a consumer's view of nature, which did not demand much reflection further than the recognition that the natural environment was undergoing a process of despoliation. In those days of relative affluence, having achieved the satisfaction of their more material needs, the middle classes had time to turn their attention to less immediate consumptions habits – e.g., the countryside.

The strength and persistence of this public concern necessarily required a response from governments and international bodies concerned to influence them. In 1968, UNESCO convened an international conference upon the use and conservation of the biosphere. It resulted in the drawing up of a four-point plan for a long-term 'Man and the Biosphere' research programme which can be outlines as follows (77):

(a) Natural environments – the description and classification of the world's ecosystems, and the measures required for their conservation.

(b) Rural environments – the study of 'domesticated' ecosystems, for example associated with agriculture and forestry, to ensure the suitability of techniques.

(c) Urban–industrial environments – the attempt to forecast and avoid harmful ecological and sociological side-effects associated with the use of technology.

(d) Pollution and related phenomena – the study of the effects of the many different sorts of pollution upon man.

In Britain, a Royal Commission on Environmental pollution was set up in 1970 under the chairmanship of Sir Eric Ashby, "to advise on matters, both national and international, concerning the pollution of the environment; on the adequacy of research in the field; and the future possibilities of danger to the environment" (78). In the first report, a general survey was made of the 'state of the natural environment' and it was concluded that although there was no cause for alarm measures it was essential to carry out research into certain areas, such as the pollution of estuaries and coastal waters. Four priorities for government action were recommended: the improvement of public water supply, the control of the disposal of toxic wastes on land, the control of dumping noxious materials at sea and the reduction of noise (79).

Finally, there was the much-publicised United Nations 'Conference on the Human Environment' in Stockholm in 1972. Amongst its lengthy list of proposals were several points which stressed the important role of science and technology in relation to the environment (80). For example:

(Item Number 18): Science and technology, as part of their contribution to economic and social development, must be applied to the identification, avoidance and control of environmental problems for the common good of mankind.

(Item Number 19): Education in environmental matters for the younger generation as well as adults . . . is essential to broaden the basis for enlightened opinion and responsible conduct . . . in protecting and improving the environment.

(Item Number 20): Scientific research and development in the context of environmental problems both national and multinational must be promoted in all countries.

Thus a series of objectives were formulated as part of, and in response to public political pressure and further research was seen as one of the chief agents for achieving them (81). The response to these goals has been the emergence of environmental science, the considerable, if temporary, growth in ecology, and research on environmentalist issues by scientists from almost every conceivable natural and social science discipline, though with a concentration, in Britain at least, in the areas of chemistry, biology and civil engineering.

From these two significantly different cases of goal establishment, it should be apparent that a variety of mechanisms and different patterns of events can lead to the establishment of goals which effectively enter cognitive structures and direct research.

5. Goal Mediation

Transformation of high level goals is necessary before they can become practically useful as guides to scientific work, i.e., before they become instrumentally potent. The resultant 'concrete' goals may have been extensively mediated by scientific and socio-economic considerations. The high level goal may still remain after this process, as a guide to further action, but it will have a changed significance as a result of its practical interpretations.

The first mediating influence will be the perceived state of the relevant science or sciences. Van den Daele *et al.* in a series of papers (82) have been concerned particularly to demonstrate that 'goal-orientation is subject to the limits set by the internal structure of the sciences'. Through the concepts of 'finalisation' and 'cognitive deficits', they have developed a useful analysis of the way in which the particular stage of development of a science may constrain the kind of goal-direction to which it is amenable.

However such an approach seems excessively deterministic, as successful goal-direction is regarded primarily as a function of the state of development of the 'relevant' knowledge base. The determination, modification and defence of goals is an essentially political process in which the knowledge is used as an ideological weapon, or as 'cultural capital' (83) to enable scientists sharing a particular world view to maintain and extend their explanatory hegemony. Thus the scientific mediation of high-level goals is anything but a passive process, the outcome determined by the state of relevant knowledge; rather it may well be marked by competition and conflict.

Such competition between scientists of different disciplinary back-grounds in the process of the scientific mediation of the goal of estab-lishing safety levels for environmental lead concentration has been clearly demonstrated by Robbins and Johnston (84). Similar conflict in the legitimation of survey methodology in the social sciences has been described by Bourdieu (85). Weingart has used the term 'relabelling' (86) to describe the response to competition for intellectual authority in which scientists merely 'dress up' their current research activities and interests in order to make them appear relevant to a newly emergent goal. Such a response was prevalent under the ecology label during the environmental movement. A significant proportion of so-called environmental science in Britain can also be described as relabelled (87). The degree of confusion, conflict and anomie arising within a formerly coherent group of scientists when faced with the development of a new goal and the requirement for some form of scientific mediation is well demonstrated by Nelkin's (88) account of the impact of the environmental movement on ecologists.

An illustration from a different field of conflict over scientific goal-mediation is provided by Hay in arguing for a different approach to the achievement of the solar energy high level goal:

. . . the solar program, through 1974, appears to have caused greater harm to the early commercialisation of solar energy research than it has been helpful to it. The basic error of federal solar energy research for the heating and cooling of buildings was its initial concept. Failure to begin a state-of-the-art study with a comprehensive analysis of working systems and field data resulted in overemphasis on mechanical (active) systems and disregard of passive and architectural systems. The consultants who were used were best known for many years of engineering research that had minimal acceptance in the market place and in filling national needs. Long-term research proposals, largely restricted to the interests and capabilities of the consultants or advisors were presented as the recommended national policy for solar energy development. Engineering theory and computer analyses were regarded as the high technology assurance of success. Institutions that depend upon endless research by people not noted for practical accomplishments were favoured with big contracts. Impressive documents and 'sideshows', which in many cases seemed motivated more for political and appropriate support than for practical results, were objectives (89).

As well as mediation in accord with scientific and technical feasibility, high level goals may be mediated also by socio-economic interests. For example, in the recent evaluation of energy programmes in the U.K. (90) it is apparent and unsurprising, that socio-economic considerations played a major role. Thus the research programmes need to be designed to achieve the goal of supplying energy at an economically competitive cost (translated in specific terms into the generation of power at £x per kilowatt equivalent by a certain date). Other more social constraints require that the energy be produced, stored and transmitted in an 'environmentally acceptable' fashion. Such factors mediate the high-level goals to produce practical goals for research. Clearly the goals of tribology and the other industrial technologies (91) established in the U.K. are strongly determined by economic goals. Indeed one of the specific claims for the establishment of tribology was the saving of £515m per annum. The socio-economic and more specifically political mediation of the goal component of the cognitive structure of the traditionally 'basic' sciences are less evident, but have been demonstrated in a number of cases (92).

The following case of solar energy research offers an excellent illustration of the scientific and socio-economic mediation of high level goals.

Solar Energy

As previously stated, it can be argued that the high level goal of solar energy research is "making the sun work for mankind by exacting large amounts of useful energy from solar radiation" (93). This high level goal has been translated into three general research goals: solar energy conversion, collection and storage, and transmission or transportation. These have been refined by socio-economic interests into three primary areas of concern: thermal energy for buildings, provision of renewable non-polluting fuel sources, and electric power generation. This combination of goals applying to particular concerns has been further mediated by scientific and technological interests to produce a series of specific research goals, including design of water heating and building heating and cooling systems, development of bioconversion and chemical reduction methods to produce fuels, and electricity generation through thermal, photovoltaic, ocean thermal difference, and wind, conversion (94). A wide range of interests have been involved in the scientific mediation, including the cognitive structures and concerns of scientists from such specialties as photochemistry, photobiology, electrochemistry, heat transfer (a sub-specialty of engineering), materials science, fluid mechanics, building science and architecture.

By this process of mediation, which, it should be noted, is neither static nor 'once and for all', a series of specific goals have emerged which have been integrated to a varying extent into the cognitive structures of the various fields (depending among other things, on 'cognitive deficits' and elite interests) and have provided one source of direction of scientific research in these various fields (95). Of course scientists in different specialties may be addressing quite different specific problems under a common goal. For example, for the electrochemist the storage goal might be resolved into a programme of research on the problems of electric battery research, whereas for the materials scientist the goal might be that of finding a compound with properties appropriate for thermal storage.

A particularly clear illustration of the operation of goal-direction is provided by the development of the sub-specialty, photovoltaic research, over the past twenty years (96). Photovoltaic research is concerned with the direct conversion of light to electricity by a quantum process that effectively avoids the severe limitations that the Second Law of Thermodynamics imposes on thermal systems. The photoelectric effect was discovered by Becquerel in 1839, but only became more than a curiosity in 1954 when three groups of U.S. industrially based scientists independently reported the development of silicon cells capable of conversion efficiencies of 6% (97). From the outset silicon was the favoured material because, though other compounds offered slightly better theoretical characteristics, silicon technology had been thoroughly developed in the electronics industry and was cheap.

The major goals for improvement of these devices were seen as increased efficiency and decreased price (98). The specific goals of the efficiency-oriented research were described as overcoming:

(1) radiant energy loss by reflection from the surface of the cell;
(2) loss of up to 30% of the hole-electron pairs by recombination;
(3) dissipation of electrical energy within the cell as heat (due to the internal resistance of the cell);
(4) the presence of a 1.1 micron wavelength energy threshold for photon absorption (99).

These goals directed photovoltaic research after 1954, and the major research was into the mechanism of the photovoltaic effect, the relation between different materials and types and levels of impurities and efficiency, and improved production techniques.

However in about 1960 these goals were almost totally replaced by the new goal, emerging from the U.S. space programme, of developing an autonomous power system suitable for space vehicles. Commitment to this new goal was so complete that little 'terrestrial' photovoltaic research continued. Moreover the new programme was remarkably successful, leading to the lightweight solar cell power systems so vital for the space programme.

In contrast to the terrestrial programme, the specific goals of space-orientated photovoltaic research were the maximisation of power and lifetime and minimisation of area and weight under conditions where cost was not paramount (or even significant). This led to quite different research, focussing on the development and assessment of special materials, the engineering of lightweight arrays and the design of extremely reliable electronic systems. Specific new areas of research included:

(i) Spectral response – because the peak in the number of absorbed photons occurs at lower wavelengths in space, junction design needs to be quite different from that for terrestrial cells, which require a lower resistance and a deeper junction.

(ii) Particle radiation – radiation damage so affected the performance of space cells that this led to the failure of some early space missions. As a consequence, understanding the nature of radiation damage and the design of radiation resistant cells became a priority which generated new theory and completely different cell designs.

(iii) Environmental protection – apart from the very low temperature, the environment of space is relatively benign for electronic devices, at least when compared with the corrosive and abrasive earth atmosphere. As a result extremely precise engineering of devices could be achieved without the need for the protective and robust features required on earth.

The most recent phase of photovoltaic research is associated with the decline of the space programme. For some period there was a failure to re-orient goals to new demands and much of the research after 1970 continued along the lines of the space model. It was only in the wake of the 'energy crisis' of 1974 and a substantial increase in the funding for photovotaic research as part of solar energy research budgets in most industrialised countries that new goals, and subsequently new lines of research, started to emerge.

6. Goal Evolution

By goal evolution we do not mean the type of radical goal replacement typified in the previous section by the sudden shift to space-oriented goals in photovoltaic research. Nor is the term evolution used with any connotation of positive progress. Rather goal formulation and reformulation is conceived of as occurring in the medium of interpenetrating systems of scientific and social production; such that subsequent statements will also involve the now 'intermediate' goal statements and to some extent incorporate them into any new statement, as well as conditioning evolving socio-economic and cognitive perceptions. In other words the interaction of the new knowledge produced under a specific goal with existing knowledge snd socio-economic conditions can lead to a redefining of this goal. It evolves in precisely the same way as other elements of the cognitive structure. Thus there is a distinct similarity between this concept of goal evolution and that of 'goal succession' developed by Sills (100) in discussing goal changes in organisations, apart from the special cognitive character of scientific goals.

Documentation of goal evolution requires extremely detailed examination of the interaction of goals, cognitive structures, socio-economic constraints and consequent knowledge over a period of time (101). Edge and Mulkay's excellent study of the emergence of radio astronomy offers indirect evidence of the operation of goal evolution.

In radio astronomy the pursuit of legitimate research goals was followed by the provision of institutional support, rapid growth, *the gradual redefinition of research goals*, . . . and eventually the extensive alteration of the framework of astronomical thought (102) (our italics).

The development of the field is portrayed in terms of "a dialectic between astrophysical (and cosmological) goals and technical means" (103), with the high level goal indicated by the statement "our ultimate goal is to find out how the sky works" (104) and specific goals suggested by such as the four aims of the galactic work in the 1951 Jodrell Bank Report (105) and the range of cosmological objectives open once the star survey phase of development was completed (106). They also note the possibility of goal evolution in Mullin's study of the phage group (107):

> Thus, although it may be true that the key problem of the phage workers was always that of "how living matter manages to record and perpetuate its experience", it may also be true that this problem was interpreted in very different ways at various points in time, giving rise to new lines of investigation (108).

Other studies which suggest the process of goal evolution include Nelkin's description of developments within ecology during the environmental movement (109) and the emergence of Verfahrenstechnik by Bucholz (110).

7. Conclusion

Ravetz (111) sees the way in which the history of science has been shaped by a faith in science as the ultimate progressive and liberating force on account of its claims to truth and objectivity as a tragedy, because the faith can no longer be justified, whether by epistemological analysis or recent human experience. But if it is tragedy for the historians of science, how much more dangerous is the retention of this nineteenth-century ideology of pure science in an age where the knowledge scientists can produce may be of vital importance to meet pressing new needs and where the competition for resources presents the science administrator with ever more difficult choices of where to concentrate scientific development. It is vital to recognise that the vast majority of science today, whether we like it or not, is 'useful science'.

There has been a considerable growth in the study of science, and undoubtedly many of the models and theories developed in the study of pure science can be adapted to useful science. However, there is a need to progress beyond this 'spin-off' mentality to direct our analysis squarely at the contemporary phenomenon. To this end, we have argued that one of the distinctive features, and a very significant one at that, of modern science, is that it is goal-directed, and only an analysis which incorporates this can hope to comprehend it. It should be noted that we are not merely claiming that "Disciplines attain phases of development in which their research can be directed to external problems" (112). Rather we are arguing that all modern science is, in principle, goal directed, so that the development of a completely general analysis is possible.

To this end the concepts of goal establishment, goal mediation and goal evolution have been articulated. These concepts, distinct only for analytical purposes, are seen as providing a basis for exploring the inter-action between scientific, technical, political, economic and social interests in the shaping of objectives. Goal establishment encompasses the role of the community of elite scientists, as one of many other lobbies, in their struggle to have their interests represented in political programmes, and of the contribution of the level of development of the knowledge-base to what is seen as practicable action. Goal mediation describes the more direct involvement of scientific interests in the mediation of political goals. Cognitive and institutional deficits (113) play an important role in this process, as do quite specific social and economic constraints. Finally, the development of the knowledge-base under goal direction, and also its impact upon social and economic problems can together or separately lead to goal evolution.

The implications of the goal-direction of contemporary science for science policy are considerable. Firstly, a recognition that the production and development of scientific knowledge *is* determined by the political interaction of socio-economic need with scientific and technical capability, and that goal-direction, occasionally explicitly but more often implicitly, is operating on contemporary science should serve to penetrate and disperse the mystification of the pure science ideology. Secondly, further examination of goal-direction may provide the basis for a more effective linking of knowledge to needs. Of course, the question of 'whose needs' remains.

Notes and References

1. J. R. Ravetz, 'Tragedy in the History of Science', in M. Teich and R. Young (eds.), *Changing Perspectives in the History of Science*, London: Heinemann, 1973, pp. 204–222.
2. *ibid.*, p. 211.

3. *ibid.*, p. 211.
4. *ibid.*, p. 212.
5. In most OECD countries, more than three-quarters and much more in certain cases, of national R & D resources are concentrated in the business and government sectors; see Y. Fabian, A. Young *et al.*, *Patterns of Resources Devoted to Research and Experimental Development in the OECD Area 1963–1971*, Paris: OECD, 1974.
6. Again in most OECD countries, R & D directed towards defence, economic, space and nuclear objectives accounts for approximately 90% of national R & D budgets; see Y. Fabian, A. Young *et al.*, 1974, *op cit.* note 5.
7. S. Blume, *Toward a Political Sociology of Science*, New York: Macmillan, 1974, Chapter 5 and H. and S. Rose (eds.), *The Political Economy of Science*, London: Macmillan, 1976.
8. G. Ciccotti, M. Cini and M. De Maria, 'The Production of Science in Advanced Capitalist Society', in H. and S. Rose, 1976, *op. cit.*, note 7, pp. 32–58.
9. *A Framework for Government Research and Development*, Cmnd., 4814, HMSO, London, 1971; for an analysis of the issues see R. Williams, 'Some Political Aspects of the Rothschild Affair', *Science Studies* 3, 1973, pp. 31–46.
10. R. Johnston, 'Contextual Knowledge: A Model for the Overthrow of the Internal/External Dichotomy in Science', *Australian and New Zealand Journal of Sociology* 12, 1976, pp. 193–203.
11. Y. Ezrahi, 'The Political Resources of American Science', *Science Studies* 1, 1971, p. 117.
12. R. Johnston, 1976, *op. cit.*, note 10.
13. A focus that has been criticised in R. Johnston and D. Robbins, 'The Development of Specialties in Industrialised Science', *Sociological Review* 25, 1977, pp. 87–108.
14. See the 'Frascati Manual', (1975) for the 'accepted' definitions within the science policy field for the various categories of research: *The Measurement of Scientific and Technical Activities: Proposed Standard Practice for Surveys of Research and experimental Development*, Paris: OECD, 1975.
15. The prolonged debate over the Rothschild Report 1971 *op. cit.*, note 9, in the U.K. can be seen as primarily over where this dividing line should be drawn and, once drawn, what form and distance of separation should be established between the entities on either side.
16. The 'science policy' boom of the 1960s, directed as it was in each country by a small group of elite scientists with experience of the science-politics interface could also be interpreted as an attempt to develop a 'scientific' basis for the management of science, and thus to ward off growing political determination of not only the level of funding, but the major directions of growth of research, cf J. Haberer, *Politics and the Community of Science*, New York: Van Nostrand, 1969.
17. R. Johnston, 1976, *op. cit.*, note 10, p. 201.
18. S. Blume, 1974, *op. cit.*, note 7.
19. B. M. Hessen, 'The Social and Economic Roots of Newton's Principia', in N. I. Bukharin *et. al.*, *Science at the Crossroads*, London: Cass, 1971.
20. J. D. Bernal, *The Freedom of Necessity*. London: Routledge and Kegan Paul, 1949 and *The Social Function of Science*, Cambridge, Mass.: MIT Press, 1960.
21. J. Needham, *Time: the Refreshing River*, London: Allen and Unwin, 1943.
22. For example P. Forman, 'Weimar Culture, Causality and Quantum Theory

1918–1927', *Historical Studies in the Physical Sciences* 3, 1971, p. 1–115, postulates the relation between the emergence of quantum physics and the hostility to strict determinism in Weimar culture; R. M. Young, 'Malthus and the Evolutionists: the Common Context of Biological and Social Theory', *Past and Present* 43, 1969, pp. 109–145 and 'Darwin's Metaphor: Does Nature Select?', *The Monist* 55, 1971, pp. 442–503 examines the social basis of Darwin's evolutionary theory; R. S. Cowan, Francis Galton's 'Statistical Ideas: the Influence of Eugenics', *Isis* 63, 1972, pp. 509–528. With regard to more contemporary science, J. Slack, 'Class Struggle Among the Molecules' in W. Pateman (ed.), *Counter Course*, Harmondsworth: Penguin, 1972, has argued that the concentration on the atypical single product, high yield chemical reaction in academic organic chemistry is a response to the needs of the chemical industry, and R. Johnston and D. Robbins, 1977, *op. cit.*, note 13, have made a general analysis of the role of occupational values in determining the development of theory and technique.

23.   G. Böhme, W. van den Daele and W. Krohn, 'Alternativen in der Wissenschaft', *Zeitshrift für Soziologie* 1, 1972, pp. 302–316; P. Weingart, 'On a Sociological Theory of Scientific Change' in R. Whitley (ed.), *Social Processes of Scientific Development*, London: Routledge and Kegan Paul, 1974; G. Böhme, W. van den Daele and W. Krohn, 'Finalization in Science', *Social Science Information* 15, 1976, pp. 307–330; W. van den Daele, W. Krohn and P. Weingart, 'Political Direction of Scientific Development' in E. Mendelsohn, P. Weingart, and R. D. Whitley, (eds.), *The Social Production of Scientific Knowledge*, Dordrecht: Reidel, 1977; for critiques see R. Johnston, 'Finalization: A New Start for Science Policy?', *Social Science Information* 15, 1976, pp. 331–336 and J. M. D. Symes, 'Policy and Maturity in Science', *Social Science Information* 15, 1976, pp. 337–347.

24.   W. van den Daele *et al.*, 1977, *op. cit.*, note 23.

25.   Case studies include Verfahrenstechnik, biotechnology, cancer research, computer science, educational research, environmental research, heavy ions research and plasma physics.

26.   R. Johnston, 1976, *op. cit.*, note 10.

27.   R. Johnston and D. Robbins, 1977, *op. cit.*, note 13 and D. Robbins and R. Johnston, 'The Role of Cognitive and Occupational Differentiation in Scientific Controversies', *Social Studies of Science* 6, 1976, pp. 349–368.

28.   J. Habermas, *Towards a Rational Society*, London: Heinemann, 1972, p. 91.

29.   R. Johnston, 1976, *op. cit.*, note 10.

30.   For example, see R. M. Cyert and J. G. March, *A Behavioural Theory of the Firm*, Englewood Cliffs, N.J.: Prentice-Hall, 1963, D. L. Sills, *The Volunteers: Means and Ends in a National Organisation*, Glencoe, Ill.: Free Press, 1957, and C. Perrow, 'The Analysis of Goals in Complex Organisations', *American Sociological Review* 26, 1961, pp. 854–866, respectively.

31.   R. Merton, *The Sociology of Science*, Chicago: University of Chicago Press, 1973, Chapter 13.

32.   L. Sklair, *Organised Knowledge*, London: Paladin, 1973, p. 66.

33.   M. Richter, *Science as a Cultural Process*, London: Muller, 1972, p. 14.

34.   J. R. Ravetz, *Scientific Knowledge and its Social Problems*, London: Oxford University Press, 1971.

35. W. O. Hagstrom, *The Scientific Community*, Carbondale, Ill.: Southern Illinois University Press, 1975.
36. *ibid.*, p. 222.
37. *ibid.*, p. 222.
38. *ibid.*, p. 180.
39. *ibid.*, p. 186.
40. *ibid.*, p. 176.
41. *ibid.*, p. 193–194.
42. *ibid.*, p. 220.
43. *ibid.*, p. 227.
44. *ibid.*, p. 163.
45. R. Merton, *Social Theory and Social Structure*, Glencoe, Ill.: Free Press, 1949, Chapter 4
46. H. A. Simon, *Administrative Behaviour*, New York: Macmillan, 1961 (2nd. ed.), pp. 63–64.
47. M. Richter, 1972, *op. cit.*, note 33, p. 15.
48. R. Whitley, 'Umbrella and Polytheistic Disciplines and their Elites', *Social Studies of Science* **6**, 1976, pp. 471–498 and C. F. A. Pantin, *On the Relations Between the Sciences*, London: Cambridge University Press, 1968.
49. J. R. Ravetz, 1971, *op. cit.*, note 34.
49a. P Bourdieu, 'The Specificity of the Scientific Field and the Social Conditions of the Progress of Reason', *Social Science Information* **14**, 1975, pp. 19–47; also R. Whitley, 1976, *op. cit.*, note 48 and 'The Structure of Scientific Disciplines and their Elites', in G. Fourez and J. F. Malherbe (eds.), *The Stakes of Scientific Professional Training*, Belgium: Presses Universities de Narmur, 1975.
50. R. D. Whitley, 'The Sociology of Scientific Work and the History of Scientific Developments', in S. Blume (ed.), *New Perspectives in the Sociology of Science*, London: Macmillan, 1977.
51. R. D. Whitley, 'Components of Scientific Activities, their Characteristics and Institutionalisation in Specialties and Research Areas: a Framework for the Comparative Analysis of Scientific Developments', in K. Knorr, H. Strasser and H. G. Zillan (eds.), *Determinants and Control of Scientific Development*, Dordrecht: Reidel, 1975, p. 41.
52. N. C. Mullins, 'The Development of a Scientific Specialty: The Phage Group and the Origins of Molecular Biology', *Minerva* **10**, 1972, p. 55.
53. T. Jagtenberg, 'Mission Orientation in Science', unpublished M.Sc. thesis, Department of Liberal Studies in Science, Manchester University, 1975.
54. R. D. Whitley, 1975, *op. cit.*, note 51.
55. N. C. Mullins, 1972, *op. cit.*, note 52, p. 52.
56. These two examples are drawn from W. van den Daele *et al.*, 1977, *op. cit.*, note 23.
57. The shaping of knowledge by these forces has been clearly demonstrated by P. Forman, 1971, *op. cit.*, note 22 and R. M. Young, 1969 and 1971, *op. cit.*, note 22, respectively.
58. W. van den Daele, *et al.*, 1977, *op. cit.*, note 23.
59. Although high level goals are themselves, hypothetically, the mediated products of other goals, needs and primary socialisation patterns, it is analytically more convenient to take them as points of departure for further mediation.

60. W. van den Daele *et al.*, 1977, *op. cit.*, note 23.
61. R. Jungk, *Brighter than a Thousand Suns*, Harmondsworth: Penguin, 1964, S. Groueff. *Manhattan Project*, Boston, Mass.: Little Brown, 1967, and many others.
62. V. van Dyke, *Pride and Power: The Rationale of the Space Program*, Urbana, Ill.: University of Illinois Press, 1964, and A Etzioni, *The Moon-Doggle*, New York: Doubleday, 1964.
63. J. Goodfield, *Cancer Under Seige*, New York: Hutchinson, 1975.
64. Much of this case-study is drawn from an unpublished report, 'A Preliminary Investigation of "Mission Oriented" Specialties with particular reference to Tribology', by B. Gillespie, Department of Liberal Studies in Science, Manchester University, 1975; a paper by C. Knowles, 'New lamps for Old', delivered at a Meeting of the BSA Sociology of Science Study Group, at Manchester in March 1976 also describes the development of tribology.
65. A. Cameron, 'Fundamental Research in an Industrial World', *Industrial Lubrication and Tribology*, 1974, pp. 60–65.
66. D. Dowson, 'Tribology', Inaugural Lecture No. 9, Leeds University, 1969.
67. Feilden Report, 'Engineering Design', London: HMSO, 1963.
68. N. J. Vig, *Science and Technology in British Politics*, London: Pergamon, 1968, pp. 62–63.
69. *ibid.*, and H. and S. Rose, *Science and Society*, Harmondsworth: Penguin, 1971.
70. Jost Report, 'Report of the Department of Education and Science Working Group on Education and Research in Lubrication', (Tribology), London: HMSO, 1966.
71. N. J. Vig, 1968, *op. cit.*, note 68, p. 184.
72. W. van den Daele *et al.*, *op. cit.*, note 23.
73. Much of this case study is drawn from A. Ellis, 'The Emergence of Mission-Oriented Specialties: the case of Environmental Science', unpublished M.Sc. thesis, Department of Liberal Studies in Science, Manchester University, 1975.
74. R. Carson, *Silent Spring*, New York: Houghton Mifflin, 1962.
75. B. Commoner, *Science and Survival*, New York: Ballantine, 1963.
76. The ecosystem is the central unit of ecological study which emphasises that no factor operates in complete independence.
77. Adapted from R. F. Dasmann, *Planet in Peril?*, Harmondsworth: Penguin, 1972.
78. 'Royal Commission on Environmental Pollution', First Report, Cmnd. 4585, London: HMSO, 1971, p. iii.
79. *ibid.*, p. 46.
80. *Declaration on the Human Environment* submitted to the Stockholm Conference, New York: U.N., 1972.
81. Thus the Presidential Address at the 135th Annual Meeting of the British Association for the Advancement of Science by Sir Kingsley Dunham concerned the role of science in explicating the environmental threats: 'Prudence demands that no effort be spared to advance those sciences which should be able to show whether the threats are real or imagined, close or distant', British Association for the Advancement of Science, London, August, 1973; A. Mok & A. Westerdiep, 'Social Influences on the Choice of Research Topics of Biologists' in R. Whitley (ed.), *Social Processes of Scientific Development*, London: Routledge and Kegan Paul, 1974, pp. 210–223, chart the influence of the environmental movement on biologists in Netherlands; it should be noted that the relation between the environmental movement and research has been shown to be exceedingly complex, cf. P. Weingart, note 86.

82. G. Böhme, *et al.*, 1976, *op. cit.*, note 23 and W. van den Daele, *et al.*, 1977, *op. cit.*, note 23.

83. P. Bourdieu, 1975, *op. cit.*, note 49a.

84. D. Robbins and R, Johnston, 1976, *op. cit.*, notw 27.

85. P. Bourdieu. 1975, *op. cit.*, note 49a.

86. P. Weingart, 'Science Policy and the Development of Science', in S. Blume (ed.).1977, *op. cit.*, note 50.

87. Thus, the Environmental Sciences Department at Lancaster University favours planetary geology and radiophysics, while the research at the School of Environmental Sciences at the University of East Anglia is dominated by disciplinary-based geology (Ellis, 1975, *op. cit.*, note 73).

88. D. Nelkin, 'Scientists and Professional Responsibility: the Experience of American Ecologists', *Social Studies of Science* 7, 1977, pp. 75–95.

89. H. Hay, 'Solar Energy – A Better Understanding', in J. Prenis, (ed.), *Energy Book 1: Natural Sources and Backyard Applications*, Philadelphia: Running Press, 1975, p. 87.

90. See for example, W. Marshall, 'Energy Research in the U.K.', *Atom*, p. 242, December, 1976, pp. 302–313.

91. The generic title Industrial Technology describes those technologies which, like Tribology, span several disciplines and subjects and which are capable of being applied to a wide range of industries, so as to bring economic benefits on a national scale, viz., terotechnology, corrosion science and materials handling.

92. See note 22.

93. T. Jagtenberg, 1975, *op. cit.*, note 53.

94. These categories have been drawn from a list of research goals and categories of funding included in NSF/NASA, 'An Assessment of Solar Energy as a National Energy Resource', B-221659, Department of Commerce, Springfield, Va. 1973, and correspond reasonably with those of International Solar Energy Society, Memorandum to the Parliamentary Select Committee on Science and Technology (Energy Resources Sub-Committee), HMSO, London, January 22, 1975, No. 156-i.

95. It should be emphasised that at any time a number of goals, some of which are complementary and some of which are conflict, may apply to a particular research field. The situation in which there is a one-to-one correspondence between goal and specialty may be very unusual.

96. Much of this data is drawn from T. Jagtenberg, 1975, *op. cit.*, note 53.

97. Chaplin, Fuller and Pearson of Bell Telephone Laboratories, Rappaport, Loferski and Lindner of RCA and Reynolds, Leies, Antes and Morainger; efficiencies have since been improved to 11–12%.

98. At 1973 prices, the cost of electricity from silicon cells is about two orders of magnitude above that of the installed cost per kilowatt of a conventional power station.

99. G. L. Pearson, 'Electricity from the Sun' in *Proceedings of World Symposium on Applied Solar Energy*, Phoenix, Arizona: Association for Applied Solar Energy, 1955.

100. D. L. Sills, 1957, *op. cit.*, note 30.

101. Unfortunately detailed studies of the 'industrial technologies' and solar energy fields has not yet reached a sufficiently advanced stage to allow their inclusion here.

102. D. O. Edge and M. J. Mulkay, *Astonomy Transformed: The Emergence of Radio Astronomy in Britain*, New York: John Wiley, 1976, p. 397.

103. *ibid.*, p. 150.

104. *ibid.*, p. 136.
105. *ibid.*, p. 145.
106. *ibid.*, p. 197.
107. N. C. Mullins, 1972, *op. cit.*, note 52.
108. D. O. Edge and M. J. Mulkay, 1976, *op. cit.*, note 102, p. 374.
109. D. Nelkin, 1977, *op. cit.*, note 88.
110. K. Buchholz, 'Verfahrenstechnik: Its Present State and Structure', delivered at PAREX Conference, Starnberg, 1974.
111. J. R. Ravetz, 1973, *op. cit.*, note 1.
112. W. van den Daele, *et al.*, 1977, *op. cit.*, note 23, p. 40.
113. *ibid.*

PART II

THE INTERRELATION BETWEEN SCIENCE AND TECHNOLOGY
A Close-up View at Determining Factors

MILLWRIGHTS AND ENGINEERS, SCIENCE, SOCIAL ROLES, AND THE EVOLUTION OF THE TURBINE IN AMERICA

EDWIN T. LAYTON, JR.

University of Minnesota

The nineteenth century was a 'Golden Age' for mechanical invention, not merely in America, but throughout the Western World. It was also a period of scientific revolution in technology. Clearly, the new scientific technology and the mechanical inventions were related, but the precise nature of the relationship has remained somewhat obscure. The old idea that technology is simply applied science is now being rejected by historians of technology. Technological advances cannot, in general, be explained by prior advances in basic science. This pattern may fit specific cases, particularly the rise of industries based upon electricity and chemistry in the latter half of the nineteenth century, but it is less satisfactory for earlier inventions (1).

In recent years historians of technology have shifted attention to a somewhat more indirect mode of interaction between science and technology. Rather than applying the results of science, technologists borrowed and adapted the methods of science to generate the knowledge and insights needed by technology (2). At its best, as in Thomas Hughes' study of Elmer Sperry, this historical approach has at least two dimensions. First, it leads to a thorough investigation of the scientific and technological background; in the case of the gyroscope, Hughes showed that Sperry had some remarkable insights into control theory before that science had been formally developed. Secondly, it encourages a concern for the socioeconomic context of technological work and for the sorts of social roles played by technologists, a tendency also illustrated by Hughes' biography of Sperry (3).

The borrowing of scientific methods, values, and social organization is

Krohn/Layton/Weingart (eds.), The Dynamics of Science and Technology.
Sociology of the Sciences, Volume II, 1978. 61–87. All Rights Reserved.
Copyright © 1978 by D. Reidel Publishing Company, Dordrecht, Holland.

clearest in the case of engineers. It is easy to demonstrate that engineers, with the aid of scientists, adapted experimental and theoretical methods from science in order to build systematic sciences in fields like hydraulics, the strength of materials, and aerodynamics. Concomitantly, engineers also adapted scientific institutions such as the scientific professional society, the research journal, and the research laboratory, along with some of the values implicit in these institutions (4).

But the case is less clear for a large number of craftsmen whose inventions are not so obviously based on science. There has been a tendency to categorize all such inventions as 'empirical'. But this is no more than a label for ignorance, since all inventions, including the most scientific, are empirical in some degree. The implication of a dichotomy between 'scientific' and 'empirical' inventions is that the latter are not scientific in any sense. No doubt many are not. It is not difficult to find examples of inventions based on insights into technique or social needs. But there remain a large and important category of inventions, mainly by craftsmen who were not thoroughly knowledgeable in the international science and engineering of their day, but whose work was in some sense scientific.

The case of the hydraulic turbine provides an example of parallel scientific traditions. French engineers and American millwrights developed quite distinct scientific approaches to the turbine in the 1820s. American engineers who began to design turbines after 1843 developed a science that represented a sort of Hegelian synthesis of the two. Both the traditions of French engineers and of American millwrights reached a culmination at about the same time. In 1827 Benoît Fourneyron invented his hydraulic turbine; in the following year two Ohio millwrights invented an improved 'reaction wheel'. Fourneyron's turbine and the Parker Brothers' reaction wheel were analogous in many respects. Indeed, many American contemporaries thought that they were identical. In the case of *Parker* vs. *Brandt* in 1850, Justice Grier, in upholding the Parkers' patent commented that

In France, M. Fourneyron received the highest honors and most liberal rewards, for intro-ducing into use this very improvement, after it had been invented in this country by the Messrs. Parker. And it was not until the circulation of Fourneyron's paper on turbines in this country, the the public attention was fairly called to the valuable improvement of the Messrs. Parker (5).

The mistaken idea that the two inventions were the same probably stemmed from the report of an investigation of the Parkers' wheel by a committee of the Franklin Institute in 1846. This report emphasized the similarities of the two devices, while minimizing the differences in design (6). Both were turbines by any definition. The term 'turbine' had been coined by Claude Burdin to categorize a class of high-speed hydraulic motors. And both 'reaction wheels' and 'reaction turbines' were smaller and faster than traditional water wheels of the same power, since the water acted simultaneously on all of the curved blades or 'buckets' rather than on just one at a time. Both could work when submerged. Both admitted water under pressure, and this pressure declined as the water transferred its energy to the wheel. Both were efficient. The Parkers' wheel probably got about 10 to 15 percent less in total efficiency than Fourneyron's turbine, but the Parkers' motors were undoubtedly much cheaper.

Despite similarities, these two inventions differed in both design and principle. These differences reflected even greater disparities between the scientific traditions of which each was an outgrowth. And the scientific traditions were, in turn, conditioned by the social roles of millwrights and engineers as well as the socio-economic contexts in which each worked.

While Fourneyron's turbine was the expression of a French engineering tradition, it had empirical roots. Indeed, both the French engineers and the American millwrights started from the 'tub wheel', a simple impulse wheel enclosed in a tub. The tub wheel was mounted with its axis in the vertical position, and like its primitive ancestor, the ancient 'Norse' wheel, it often operated without gearing. It was very cheap, but also very inefficient. By 1620 French millwrights had converted the tub wheel into a simple reaction wheel by curving the blades so that the water was discharged in a direction opposite to the motion of the wheel (7). American millwrights made essentially the same improvement in the tub wheel sometime prior to 1795. In that year Oliver Evans published a picture of a tub wheel. And while Evans described it as operating solely by 'percussion', his illustration shows a wheel with blades curved to discharge the water in a direction nearly opposite to that of the wheel (8).

French engineers became interested in horizontal wheels with curved buckets through Bernard de Bélidore's description, particularly his illustration of the gracefully curved wheel of Basacle (9). Jean-Charles de Borda made a theoretical investigation of horizontal as well as conventional wheels, which opened a new chapter in hydraulic science (10). This line of investigation, continued by Lazare Carnot, Louis Navier, Claude Burdin, and Jean-Victor Poncelet, culminated in Fourneyron's turbine (11). This French tradition reflected the close institutional ties between science and engineering that had developed in France. French engineers were included in the Royal Academy of Sciences from its formation, and the French developed college training for engineers in the eighteenth century, culminating with the foundation of the *Ecole Polytechnique*. Except for Fourneyron, all of the French pioneers in the development of the science of the turbine were members of an academic and scientific elite. But Fourneyron, a graduate of the *Ecole des Mines*, also reflected the same scientific tradition. It was characterized by an elegant mathematical and deductive approach which paralleled the contemporary French style of physical science.

The fact that French engineers were oriented to an academic and scientific reference group was a source of great strength and accounts in good measure for Fourneyron's ultimate success. But it was not without its limitations; considerations of academic and scientific respectability could lead to the neglect of factors of importance to technology. A rigorous mathematical treatment was of necessity highly idealized. It neglected those effects arising from the internal motions of water, such as eddying and turbulence, since there was no satisfactory mathematical method of handling such phenomena until the rise of the science of fluid mechanics in the twentieth century.

The French engineers developed a new theoretical approach to prime movers. It may be seen as a small skirmish in the long debate over *vis viva*. Borda used 'living force' or *vis viva* in his analysis of water wheels, and he was able to correct a serious error by Parent. But the energy approach led him to a new way of looking at hydraulic prime movers. Borda expressed somewhat obscurely the doctrine clearly enunciated by Carnot later: Complete efficiency could be achieved if the water entered the wheel without losing any of its energy, that is without 'shock', and if, in addition, the water left the wheel without velocity, that is after all of its energy had been transferred to the machine (12). In effect, this approach led to treating the machine as a 'black box'. It concentrated on the conditions of entrance and exit. It was also highly idealized, in that it assumed that friction and water resistance could be neglected. But its advantage was that it provided the basis for a very elegant and general mathematical treatment of hydraulic machinery.

Fourneyron's turbine represented an application and extension of the French energy approach. Shockless entry could be approximated by having the water enter tangent to the surface of the blade of the wheel. All of the energy of the water could be transferred to the machine by the principle of reaction if the water were to exit in a direction exactly opposite to that of the wheel itself and without any absolute velocity. Fourneyron's achievement was to develop mathematical expressions for both of these conditions and to show how they might be approximated in an actual design (13).

It will be helpful if we treat the second of these conditions first. Navier showed how the existing reaction wheel could be improved, but as Borda had seen, a complete conversion of energy by reaction was impossible in a modified tub wheel (14). The problem was inherent in the basic design. The water would have to exit in a horizontal plane for maximum efficiency. But if the buckets were bent around to the horizontal either they or the water would impinge on the succeeding buckets. Fourneyron saw that the way around this dilemma was to curve the buckets away from the axis of the wheel, so that the water moved outward in a radial direction.

It is interesting to note that American millwrights had already made this change. Though we know very little about the early evolution of the reaction wheel in America, the standard, or most approved form in 1827, when the Parker Brothers began their experiments, was also a radial, outward-flow wheel (15).

While American and French developments were parallel concerning the form of wheel, the two traditions differed on the conditions of the water's entrance into the wheel. Fourneyron showed mathematically how to achieve tangent entry. The essence of the problem was that the effective angle of tangent entry varied with the speed of the wheel; Fourneyron showed how this angle varied with the velocity. Fourneyron derived a simple mathematical function to determine the angle of tangent entry once the speed was known. Of equal importance, he showed how the water could be directed into the wheel at a precisely pre-determined angle. His invention was fixed metal guide vanes mounted inside the wheel which channeled the water into the wheel at the desired angle (16).

While French engineers strove to achieve tangent or 'shockless' entry. American millwrights were following a different path. They thought of reaction wheels as operating by impulse as well as reaction. In the common reaction wheel, whether of the tub-type or a radial-outflow wheel, the wheel was mounted at the bottom of a reservoir or 'penstock' of water. Manifestly, the weight of the column of water imparted an impulse to the wheel at its entrance. One of the earliest and most successful American reaction wheels was intended to maximize this impulse. Benjamin Tyler's 'Wry-Fly' reaction wheel, invented in 1798 and patented in 1804, closed off the top of the 'tub' with a circular metal plate, and the water was admitted by multiple shutes which produced jets of water. But, unlike the shute in Navier's improved reaction wheel, the jets were not directed in a direction tangent to the blade, but rather at an angle calculated to increase the impulsive action of the water at entrance (17).

In the same year that Fourneyron invented his turbine, the Parker Brothers in America made a key observation. In 1827 they were building a reaction wheel for a grist mill. In Zebulon Parker's historical account, written in 1851, he recalled that

The water in consequence of the direction of its approach, fell into the wheel with a vertical or whirling motion, which was in the direction contrary to the motion of the wheel. A plank by accident fell into the penstock, and lodged in a position which changed the direction of the water so as to cause it to enter with a circular motion, coinciding with the wheel's motion (18).

Apart from its apparently accidental and empirical origin, the Parkers' invention appears to be quite the opposite of Fourneyron's.

Fortunately, in a subsequent patent suit the Parkers made a drawing in which they reconstructed the exact circumstances of their discovery. The illustration makes it clear that their accidental discovery was, in fact, a fixed guide. That is, the position of the plank was such that it served as a guide vane analogous to those used by Fourneyron (19). And, indeed, in subsequent experiments they introduced several such planks or guides, placed in various positions (20). In short, they started with Fourneyron's solution, but they did not work backward to his conceptualization of the problem. Rather, they interpreted their experiments in terms of spiral motion and impulsive action or 'percussion' of particles of water imping-ing upon the buckets. This led them to invent not fixed guides, but a spiral casing or scroll case, something not at all implicit in their original set up. The scroll, now usually called a 'volute chamber', is a casing in approxi-mately the form of an Archimedian spiral, which imparts a circular motion to the water prior to its entrance.

The Parkers' invention depended not simply on an accident, but even more importantly, upon a conceptual framework and experimental approach which was an outgrowth of a scientific tradition developed by American millwrights. This tradition was the natural outgrowth of the craft knowledge transmitted by oral tradition from master to apprentice. But by the latter eighteenth century millwrights began to accumulate a written, scientific literature. Oliver Evans expressed this new spirit in America. His *The Young Mill-Wright and Miller's Guide*, published in 1795, was consciously scientific in spirit. Nearly half of the book was an elementary scientific survey of the principles of mechanics and hydraulics (21). Restricted to elementary mathematics, the mechanics was limited to rather simple cases. But the rules of hydraulics provided means for calculating the most important parameters, such as velocity, quantity of flow and pressure, and enabled millwrights to predict how these would be modified by altering the dimensions of the channel through which the water moved. Thus, of necessity, millwrights had to proceed by tracing the water's path as it moved through their machines.

While hydraulics provided a body of quantitative theory, Evans warned that it had severe limitations. He noted that

in treating of the elementary principle of hydraulics, it is necessary to proceed upon theoretical principles; but let it always be recollected that from various causes . . . and particularly from that essential property in them, the perfect mobility of their particles among each other, the phenomena actually exhibited in nature, or in the processes of art, in which the motion of water is concerned, deviate so very considerably from the deductions of theory, that the latter must be considered as a very imperfect guide to the practical mill-wright and engineer (22).

And Evans concluded that theory had to be constantly checked by accurate experiment. In this John Smeaton, the British civil engineer, provided a model not only for Evans, but for his readers; Evans so admired Smeaton's work that he incorporated the full text of Smeaton's classic study of hydraulic prime movers into his book, along with his own critical examination of Smeaton's and other current hydraulic experiments (23). Evans presented a set of rules for applying scientific methods to technology, which included the discovery of fundamental principles, making deductions from them, and testing these results by experiment. In each case he illustrated these steps with examples drawn from his own career as inventor (24). In short, Evans' science was practical, hydraulic, and experimental.

The Parker Brothers were intimately acquainted with Evans' work, and its influence can be seen in their theoretical and experimental work. The mechanics provided a conceptual framework which was of enormous value. Zebulon Parker reasoned from Newton's laws of motion, and though hampered by the variations in phrasing found in Evans and other second-hand accounts, he was led to correct conclusions on the nature of reaction wheels and how they worked (25). But their science was primarily hydraulic. They supplemented Evans with Zachariah Allen's *The Science of Mechanics*, published in 1829, and other works (26). They sometimes formulated problems in qualitative mechanical terms, then restated them quantitatively in hydraulic terms, and tested the result by experiment. Thus, after their accidental discovery of the effects of a fixed guide, they sought a general qualitative understanding in terms of the motion of particles and their 'percussion' or 'reaction' according to Newton's laws of motion. Then they translated these general principles into quantitative hydraulic terms in designing their scroll case which they then tested exhaustively to find the proportions required for maximum benefit.

The millwrights' science had severe limitations, but it was not without its strengths. The lack of a scientific reference group imposed handicaps. Millwrights had no established channels of communication with each other and with scientists. As a result they lacked consensus on many fundamentals, since there was no regular means for separating truth from error. Without knowledge of what had already been done, the Parker Brothers struggled for years to perfect a means to measure precisely the power output of water wheels (27). They were unaware of the fact that just such a device, the Prony Dynamometer, had already been invented in France. But it is significant that within less than a year after the first published American description of this device in 1843, Zebulon Parker had not only built a dynamometer and model testing flume, but had introduced a significant improvement, which considerably improved its accuracy (28). The Parkers, however, were successful in devising improved methods for measuring the quantity of water used by their wheels, which were adopted by the Franklin Institute in its own testing. The Parkers had a talent for experiment, a trait that carried them far. And the lack of a scientific reference group was not without advantages. The Parkers were not constrained to an unrealistic idealization of the phenomena. They grappled with friction and fluid motion uninhibited by the absence of adequate foundations in scientific theory. In dealing with problems of fluid behavior, they placed great emphasis upon direct visual and tactile observation of the flow of water. Thus, Zebulon Parker devised a glass-walled testing flume sometime prior to 1849, and by means of small pellets in the water observed its behavior as it passed through his machine. (29).

The nature of their social role placed limitations on the millwrights' ability to absorb and use science. They were organized on a craft basis but little changed since the Middle Ages. Apprenticeship training was a hit-and-miss affair, and the results were variable. An Ohio millwright, Conington W. Searle, was probably typical. In 1843 he testified that "I am acquainted with hydraulics and hydrostatics", but also that "my knowledge of Hydraulics is a little from theory only, and what I have picked up from observations" (30). The millwright had to be something of a jack-of-all trades. He contracted to design and build mills or parts of mills, and he did most of his own work with his own hands. Thus, specialization and study were difficult. Most millwrights built mills on smaller streams and tributaries in the hinterland; few had access to the great libraries of the major cities. The further to compound matters, millwrights were nomadic. There were few districts with enough mills to keep a significant number of millwrights permanently employed. Austin Parker traveled about Northern Ohio on what he called his 'circuit' (31). The Parker Brothers, who were among the most scientific and successful of American millwrights, were handicapped by their lack of access to books and periodicals, their sketchy training, especially in mathematics, the diversity of tasks which they performed, and by the necessity of moving from place to place in search of work.

But despite these handicaps, the Parker Brothers managed to get a wide acquaintance with a miscellaneous literature in English of science and mechanical improvement. Both followed intensive courses of self-study from an early age. But with little mathematics and no foreign languages or access to scientific libraries, neither was in touch with the international engineering literature in hydraulics. David Brelesford was a blacksmith neighbor of the Parker Brothers who knew them as boys, Testifying in 1849, he recalled that

They were both, almost constantly engaged in mechanical pursuits and studies. Zebulon particularly was always planning and studying some mechanical invention, and particularly relating to water wheels . . . they were both always considered mechanical giniuses [sic] (32).

Another neighbor, John Hearsham, a farmer, also had known the Parker Brothers since their boyhood. He remembered that

they were both always experimenting on water wheels, except they were asleep, and [I] expect they drempt about it. When they were boys they used to start little wheels & etc. on runs and brooks . . . . Their habits were to study and practice all kinds of mechanical inventions – mostly water wheels. But I have known them in the winter seasons to invent and make clocks, fiddles, thrashing machines, churns or various other things that could be coungered out of wood. They were both remarkable for their mechanical genius and study. (33).

But perhaps the greatest obstacle to the Parker Brothers' scientific work was that of finding time and money to support their research. Shortly after their invention of the scroll case in 1828 and 1829, they had to separate, apparently because there was not enough work to support both in one district. Austin's letters for the period 1829 to 1834 are filled with accounts of experiments and projected experiments. But to make a living he had to build mills. He took off time from work for his experiments, but his expectation that he might defray the costs of these interludes from patent royalties proved largely delusive, though the sums he did collect no doubt helped (34).

Austin Parker was evidently a very creative type whose mind bubbled with new ideas, not merely respecting water wheels, but steam power, with which he was fascinated, and other mechanical inventions. He concluded that "it appears to me that I must evidently be a constant slave to the genius of improvement and must begin an arrangement" (35). His favorite plan was to get support from the state legislature. In 1829 he wrote, with premature optimism, that "the legislature would be glad to assist in the undertaking . . . . The right manner of grinding with stones is not yet known and would be easily discovered by Experiments. We can try the exact power of all kinds of wheels and find the exact maximum of each kind to a fraction" (36). Another scheme was to set up a small machine shop which he hoped would provide a living while leaving enough time for scientific investigation. A third plan involved getting support from a wealthy miller in his home district. But none of these schemes came to anything, and Austin, like his brother, had to squeeze in whatever experiments he could whenever he could at his own expense. Sadly, he never had the opportunity to experiment with his ideas in steam power. But in these years he did manage to redesign their reaction wheel and to make a fundamental invention, the draft tube.

While Austin Parker never found the time and resources to pursue his interest in steam, his invention of the draft tube was, in part, inspired by his interest in the steam engine. It was also a continuation of his efforts to improve the reaction wheel. The Parker Brothers' wheels were intended for high-speed operation, particularly in sawmills, where high speed permitted the wheel to be connected directly with the sawing machinery by a simple crank and connecting rod, thus eliminating gearing and attendant friction. To achieve the necessary speed the Parkers used even multiples of small wheels mounted on a common horizontal shaft. The speed of a wheel being in inverse proportion to its size, a grouping of two, four, or six smaller wheels would rotate much faster than a single larger wheel with the same power, while the horizontal mounting allowed the shaft to be connected directly to the saw machinery. But this arrangement created a difficulty; the crank which communicated the power of the shaft had to be at the same level as the wheels. But if mounted at water level in the tail race, the crank would hit the water in its rotation; to avoid this the Parkers had to mount their wheels higher than the discharged water, thus sacrificing some of the head of water available. The draft tube was an airtight casing which allowed the reaction wheels to be mounted at a convenient distance above the tail race, without sacrificing head. The suction of the column of water below the wheel acted as effective head and was fully equivalent in effect to water above the wheels (37).

The connection between the draft tube and the steam engine appears in the phrasing used in Austin's letters to his brother. The first glimmerings appeared in a letter written on 4 May, 1831

I would like to know how fast water could be raised by the simple precept of producing a vacuity by the friction of water with air; likewise how well a wheel could be made to run in the same way . . . by the weight of atmosphere (38).

In a letter of 28 August, 1831, he proposed to build a mill that "would go by the weight of the atmosphere. The water will be suspended below the wheel in a subterraneous trunk, and the wheel run in an entire vacuum" (39). The emphasis upon creating a vacuum and working by atmospheric pressure is highly reminiscent of the operation of a condensing steam engine. Conversely, this phrasing is otherwise rather strange in the context of hydraulics, and it was abandoned in subsequent descriptions of the draft tube.

While they were making their inventions, the Parker Brothers were also striving to gain a better theoretical understanding. It is a tribute to their experimental approach that their insights were at least parallel with those of the French engineering school. This fact is all the more remarkable in that the Parkers lacked the framework of energy concepts within which to place their insights. From 1829 to 1833 Austin Parker redesigned their reaction wheel in a way that approximated 'shockless' entry. He discovered that their wheel lost power through the impact of the water, particularly with the tip or leading edge of the blade, and that there was not enough pressure (or energy exchange) between the water and the curved path of the buckets beyond the entrance. To remedy this Austin increased the size of the inlets so that the water entered the wheel at a speed only one-quarter greater than the speed of the wheel. He also bent the tips of the buckets inward to form 'wings' in order to lessen losses through the impact of the water on the tips of the blades. In effect what Austin Parker did was to reduce shock and turbulence at entrance (40).

After Austin's death in 1834, his brother Zebulon continued his experiments. The redesigned wheel was a great success According to

Ellwood Morris, an engineer affiliated with the Franklin Institute, the common reaction wheel seldom utilized more than 40 percent of the energy of the water (41). The introduction of the scroll case along with the subsequent improvements in the design of the Parkers' reaction wheel produced a dramatic improvement. When tested by the Franklin Institute in 1846 three of Parker's wheels gave efficiencies of 65, 61, and 67 percent (42). That is, they were approximately as efficient as the best breast wheels, but considerably less expensive. The combination of cheapness with efficiency led to the rapid adoption of the Parker Brothers' wheels as well as others which incorporated their improvements. In 1851 *Scientific American* editorialized that "the subject of reaction wheels is more interesting than any other, because there are twenty of such wheels used in America for one overshot or breast wheel, we suppose" (43). By 1841 Parker had gone into business manufacturing his wheels in the Ohio area, and he had licensed five other companies to manufacture them in other regions. As of 1841 he reported 800 of his wheels in operation (44).

But the redesign of their reaction wheel by the Parker Brothers posed theoretical problems. Austin had been somewhat uneasy over his own empirical findings, since having the water enter at nearly the speed of the wheel and reshaping the buckets appeared to him to sacrifice "part of the purchase of the percussion" (45). Zebulon continued Austin's work. He found that for most efficient operation the wheel should move faster than the entering water, thus virtually eliminating impulse. In so doing, Zebulon Parker further approached 'shockless entry' though he lacked the concept and its corollary of tangent entry (46). He then undertook a series of experiments, particularly one in 1841 that he called "the beautiful and interesting experiment which manifests the law of the centrifuge and equilibrium of a mass of water . . . revolving within a vertical cylinder" (47). Parker noted that "it is obvious that the simple percussion must cease as the wheel acquires a velocity greater than the water within it" (48). But Parker's experiments showed him that "the amount of forces tending to propel the wheel are not lessened because as the percussion diminishes an equivalent is produced by the centrifuge" (49). In effect what Parker demonstrated was the interconvertability of changes in velocity, that is, in absolute kinetic energy, and centrifugal effects. Parker saw that "the percussive or centrifugal force of the water is united with the reacting force into one impulse" (50). In this he approached, by a different path, a truth expressed in the fundamental Euler equation for the turbine. This equation states that the total energy is the sum of the absolute change in kinetic energy and the centrifugal effects (51).

But reasoning from Newton's laws of motion Zebulon Parker had also come to grasp the conditions of exit required for maximum efficiency. Indeed, in 1841 he claimed, somewhat optimistically, that in his wheels

the wheel's velocity at its verge . . . will have a velocity equal to the relative velocity of the water through the issues in a contrary direction, the water at the instant of its discharge, has no velocity except the radial component of its divergency from a tangent to the verge of the wheel; which is necessary to give place to the successive flow through the wheel (52).

Despite some crudities and imperfections, the Parkers' scientific achievements were impressive. The practical influence of their investigations, however, is conjectural. While their inventions had an important influence upon both millwrights and engineers alike, their experimental work was not published. Zebulon Parker submitted experimental results to the Franklin Institute in 1841 and 1844, but they were not published. *Scientific American* published a summary of this work with a few brief excerpts in the early 1850s, but in such an abbreviated form as to convey only certain conclusions without the supporting data that would carry conviction. In any case, in 1855 a new chapter in turbine science began in America with the publication of the *Lowell Hydraulic Experiments* of James B. Francis. Francis became the spokesman for an American engineering tradition, one that combined elements from both the French engineering approach and that of American millwrights. One of the principal results of this synthesis was the evolution of the amazingly versatile and economical Francis turbine equipped with a scroll case and draft tube. It was in many respects the legitimate offspring of a marriage between two traditions: that of the millwrights and that of the engineers.

*   *   *   *

Hydraulic engineers appeared in America in the 1820s and 1830s to meet the technical needs of an industrial economy. The pioneers in this industrial development were a group of Boston capitalists led by Francis

C. Lowell, who founded a textile factory complex at Waltham, Massachusetts after the War of 1812. But the power needs of the emerging factory economy vastly transcended anything heretofor encountered in America. These Boston capitalists then proceeded to harness the water power of a major river, the Merrimack, and to found the new factory city of Lowell. By 1858, the factories at Lowell used 12,000 horsepower, chiefly in the manufacture of cotton textiles. The success of Lowell led to the creation of additional factory towns tapping the power of the Merrimack at Lawrence, Nashua, and Manchester, as well as to large scale hydraulic works on other major rivers. Hydraulic engineers were needed to design, construct, and manage these elaborate and complex hydraulic systems. James B. Francis, the chief engineer and manager (or 'Agent') of the hydraulic works at Lowell from 1834 to 1885, also served as technical consultant in engineering matters to all of the factories using water at Lowell. This involved Francis and other engineers in the design of hydraulic prime movers for the factories (53).

Economic factors produced a set of constraints for the work of engineers like Francis that were quite different from those of the millwrights. Apart from the sheer size and complexity of the works which they superintended, the conditions of their work differed in two vital respects. First, the water power available at a give site was of critical concern; once all the water was committed, any expansion in a given factory meant getting higher efficiency from the fixed amount of water power available. Furthermore, the cost of the water power was high. At Lowell, the water was measured in millpowers. The initial lease on a millpower was usually $10,000 with a yearly rental of $300. Francis computed the annual cost of one millpower, allowing a 6 percent interest charge for the initial investment, at $900. At 70 percent efficiency one millpower would yield approximately 60 Horsepower (54). Under these circumstances, there was great incentive to utilize the water efficiently. On the other hand, the capital cost of the prime mover was less significant. The Lowell capitalists had invested a total over $12,000,000 by 1840, and they normally kept about one-third of their capital for working capital. Thus, engineers like Francis served a very different market than did the millwrights. There was little need to save on capital and an enormous advantage in achieving high efficiencies in the use of water power.

Millwrights built relatively small grist and saw mills serving the needs of agricultural districts. Their clients were usually small businessmen with modest capital. On the other hand, water power was usually abundant. James B. Francis well understood the differing economic constraints. He noted that

Much prejudice existed here, as elsewhere against the reaction wheels; a great number of which had, however, been used throughout the country; in the smaller mills, and with great advantage; for although they usually gave a very small effect in proportion to the quantity of water expended, their cheapness, the small space required for them, their greater velocity . . . and not requiring expensive wheelpits of masonry, were very important considerations, and in a country where water power is so much more abundant than capital, the economy of money was generally of greater importance than the saving of water (55).

But Francis found no science in the millwrights' work, and he was unaware of the work of the Parker Brothers. He noted only that "a vast amount of ingenuity has been expended by intelligent millwrights, on these wheels", and that "within a few years there has been a manifest improvement in them, and there are now several varieties in use . . . giving a useful effect approaching 60 percent" (56).

Fourneyron's turbine was imported to America in 1843 by Ellwood Morris, a Philadelphia engineer affiliated with the Franklin Institute (57). But Morris's turbines were not very successful. However, the description of his own and Fourneyron's work which he published in the *Journal of the Franklin Institute* in 1843 caught the attention of Uriah Boyden, a consulting engineer who worked at Boston and Lowell. Boyden invented an improved version of the Fourneyron turbine, and his first turbine was installed at one of the Lowell textile factories in 1844–1845 (58). Boyden's turbines got unheard of efficiencies, generally from 82 to 88 percent. In contrast, the breast wheel, at best, was able to utilize only two-thirds of the power of the water, and in practice a great many of the breast wheels at Lowell and elsewhere got efficiencies of only about 60 percent. But Boyden's turbines were very expensive. According to Francis, Boyden's early turbines were built "almost regardless of cost" (59). They required large masonry wheel pits. Boyden insisted on making all the internal surfaces smooth. This involved using expensive imported Russian sheet iron for the buckets and, in some cases, bronze casting for the guides. Boyden eliminated all internal projections such as bolts, which also added to the cost. Boyden's early turbines cost well over $100 per horsepower. In 1847 the Proprietors of Locks and Canals which owned and managed the water power at Lowell purchased the local rights to Boyden's turbines and they produced Boyden turbines in their own machine shop. By 1858, Francis reported that the average cost per horsepower was approximately $77 (60). There is no comparable cost data for reaction wheels. But it was universally conceded that the millwrights' reaction wheels were much cheaper, probably in the range of $20 to $30 per horsepower.

The American engineering tradition in turbine science resembled a synthesis of those of French engineers and American millwrights. Both Boyden and Francis were intimately acquainted with the European writings on hydraulics and the turbine. And there is no doubt that they benefited from them. Their private correspondence reveals that they absorbed many specific results, such as the methods for calculating tangent entry and exit without velocity. It is, therefore, surprising to find both men rejecting not only the continental tradition, but virtually all mathematical theory in hydraulics. Francis commented that

the turbine has been an object of deep interest to many learned mathematicians, but up to this time, the results of their investigations . . . have afforded but little aid to the hydraulic engineer (61).

In a letter to an aspiring theoretician Francis held that "little of what we call the science of hydraulics has any practical value except that founded directly on experiments". And Francis confessed, "I fear I have a habit of doubting purely theoretical deductions in an unreasonable degree" (62). Similarly, in an unpublished article on the history of the turbine, Boyden denounced the mathematical approach which he thought had led Euler and Newton into 'great absurdities' (63).

In actual fact both Boyden and Francis borrowed extensively from the theoretical literature in hydraulics. Their correspondence often involved discussions of mathematical theory, and the European literature underlay much of their thinking. Indeed, Boyden wrote to Joseph Henry of the Smithsonian that most of his discoveries were mathematical in character (64).

Their public rejection of mathematical theory appears to have been a compound of philosophical conviction and practical necessity. Both

Boyden and Francis had an affinity for an Anglo-Saxon empirical tradition, and their attitudes may have reflected national styles in science. At several points in their correspondence Boyden admonished Francis on idealization, though both men treated it as a practical rather than a philosophical matter (65). Certainly, idealization involved assumptions which were based on mathematical necessity rather than physical reality. The theory assumed no friction, laminar flow (no eddying or turbulence), completely filled water channels, operation at full gate (or maximum power), the absence of shock at entrance (given certain mathematical conditions), uniform motion of the water, uniform angular velocity of the wheel, and constant internal pressure as the water moved from the guides to the wheel. (66). None of these corresponded precisely to reality. More to the point, given Fourneyron's turbine, further improvements in efficiency depended upon addressing these omitted factors.

Fourneyron never published his method for determining the speed of maximum efficiency for his wheel. Since the theory presupposed that the speed was known, this produced a situation a bit like the fable of the emperor without his clothes. The modern, exact equation for the 'specific speed' of turbines depends upon considerations of similitude and dimensional analysis that were not available until the late nineteenth century. But in addition, the speed behavior depends on the geometry of the machine; thus, each design of turbine has its own 'specific speed' (67). In his *Lowell Hydraulic Experiments*, Francis presented a number of design principles including means for calculating the speed, but these rules were explicitly based on empirical data (68).

Indeed, the whole idea of a pure reaction turbine with fixed guides was a mathematical fiction. When less than full power was needed, the gate admitting the water was partially closed. But at part gate the efficiency of Fourneyron's turbine fell off drastically. Francis included in his book a characteristic curve which showed just how poor the performance was at part gate (69). The problem lay in the fact that at part gate the speed changed and with it the angle of tangent entry. Thus, in actual practice turbines with fixed guides operated by impulse a good deal of the time, and shock at entrance was by no means eliminated.

In attempting to come to grips with the factors omitted, of necessity, in the mathematical theory, Boyden and Francis were led to an approach that was essentially the same as that adopted earlier by American mill-wrights. For the engineers it was a matter of conscious choice; for mill-wrights such as the Parkers, it had been a matter of necessity. Nor is there any evidence that Boyden or Francis borrowed theory from the mill-wrights; they were unaware of the work of the Parker Brothers, since the Franklin Institute had refused to publish Zebulon Parker's treatise. But having rejected classical mechanics, Boyden and Francis were driven to adopt a hydraulic approach. Thus, like the millwrights, they proceeded by a detailed analysis of the actual path of the water. They attempted, by whatever means they could find, to reduce friction, eddying, and turbu-lence, even in the absence of an adequate scientific understanding of these phenomena. And, skeptical of theory, both Boyden and Francis put great emphasis upon experiment. These similarities, particularly the absence of abstruse mathematics, made the *Lowell Hydraulic Experiments* accessible to millwrights and facilitated a creative interaction between millwrights and engineers.

There were many parallels between the works of American engineers and millwrights. Both were concerned with reducing friction. That had been of the essence in the Parkers' original invention: the elimination of gearing by the use of even multiples of small wheels on a horizontal shaft. Boyden, like the Parkers, sought to reduce friction. After trying a number of expedients, he invented a low-friction bearing upon which to suspend his turbines, consisting of a series of rings on the shaft which fitted into grooves in the bearing. This gave a much larger bearing surface, so that each square inch had to bear only 46 pounds though the total weight of the machine might be seven or eight tons. Moreover, Boyden coated the bearing surfaces with a low friction-metal (70).

On a more fundamental level, both American millwrights and engineers were concerned with maintaining the smooth flow of water and pre-venting, in so far as possible, all turbulence or eddying. To guide the water gently and without shock into the fixed guides, Boyden adopted a variation of the Parkers' scroll case (71). Francis later abandoned this, probably for reasons of convenience and economy, but the scroll or volute chamber has become subsequently a standard part of the setting of the Francis turbine.

Exit without velocity was another mathematical fiction not realized in practice. To reclaim part of the kinetic energy of the discharged water, Boyden invented the diffuser. It consisted of two flared, circular collars through which the water flowed as it left the turbine (72). Since the path was diverging, the water was slowed down in conformity with the Venturi principle. This had the effect of transferring some of the excess kinetic energy back to the turbine. Francis later investigated the principles of Venturi tubes and published his results in the second edition of the *Lowell Hydraulic Experiments* (73).

The diffuser in its original version was excessively expensive, since apart from its own cost it required larger wheel pits. Francis omitted diffusers in many later installations. The gain in efficiency was not worth the added cost; Boyden's diffusers reclaimed only 3 percent of the 5 percent theoretically possible. But the diffuser or Venturi principle could be combined with the draft tube by giving the tube an outward taper in the form of an inverted truncated cone. Curiously, neither Boyden nor Francis saw this possibility. But in one of the fragments of Zebulon Parker's lost treatise published in *Scientific American* in 1851, there is an illustration of a draft tube with a distinct outward taper at its lower end (74). This appears to be the first attempt to combine the draft tube with the diffuser principle. The idea was revived in the 1880s for inward-flow turbines, and the tapered draft tube, like the scroll, has become a standard part of the setting of Francis turbines.

Both Boyden and Francis gave great attention to precise experimental testing of turbines. In this they were very successful, and their work became the basis for systematic industrial research in turbine design in America. But here, too, their work was parallel to that of the Parker Brothers. Both Boyden and Zebulon Parker improved the Prony Dynamometer; indeed, they used the same improvement, the dash pot. The Dynamometer worked by applying a massive wooden friction brake to a wheel mounted on the shaft of the turbine. But accurate measurement of the turbine's torque was made difficult by the oscillations produced by the intense friction of the brake on the wheel. The dash pot dampened these oscillations; it consisted of a piston immersed in a cylinder of water attached to the arm of the brake. It is quite possible that Boyden simply borrowed this innovation from Parker who used it in 1844, months before Boyden tested his first turbine in 1845. Parker's testing apparatus and results were used as evidence in the case of *Parker vs. Hatfield* in 1844 (75). There is every reason to suppose that Boyden, who made a considerable fortune on his patents, followed patent litigation with great attention.

The Prony Dynamometer measured the output of a turbine; to measure the input required an exact calculation of the amount of water used. Francis, with some mathematical assistance from Boyden, derived a new formula for measuring the quantity of water flowing over weirs, and this became one of the fundamentals of turbine testing in America (76). Here, too, Zebulon Parker's work was parallel. Parker invented an improvement in the traditional millwright's method of measuring the flow of water (77). But he was apparently not satisfied. In 1850 he invented a water meter intended to measure the flow of water with minimum disturbance. It was unpatentable, since it represented an adaptation to this new purpose of a well-known design for a rotary steam engine. It appears likely that Parker may have intended his water meter as a means of measuring the flow of water into turbines, though there is no evidence that this device was ever used for any purpose (78).

The greatest achievement of the American tradition was the Francis turbine. I do not have the space to reconstruct here the complex evolution of this remarkable and versatile device. But one point may be emphasized; the Francis turbine was the result of interactions between American millwrights and engineers. The Francis turbine is sometimes called the 'mixed-flow' turbine, since the water moves both inward and downward. The first inward-flow turbine was invented by a millwright, Samuel Howd of Geneva, New York. Both Francis and Boyden at about the same time realized that the inward-flow principle could be used to obviate a difficulty inherent in the outward-flow design. The diverging passages of an outward-flow wheel inevitably produced spreading and eddying. A simple inward-flow turbine such as Howd's was inherently inefficient, since the converging passages accelerated the water, which left the wheel with considerable velocity. Boyden and Francis routinely calculated the cross sectional area of the water's path, inch-by-inch as it flowed through the turbine. This approach led to a solution: to compensate for the narrower width by increasing the depth of the channel, thus giving the water a downward as well as inward motion (79).

Both Boyden and Francis were disappointed by their designs for an inward-flow turbine. Due to several defects, Francis' version gave less efficiency than Boyden's improvement of Fourneyron's design (80). But several millwrights saw potentials in the design that Boyden and Francis had missed. At about the same time. Asa M. Swain and James Leffel realized that by extending the buckets further downward, the capacity of the turbine could be radically increased. Their work was further extended by another millwright, John B. McCormick in 1876 (81). Increasing the capacity of the turbine had enormous economic results; since the quantity of water processed increased, turbines of a given size and cost could be made more powerful. Thus, for a turbine of 36 inches in diameter operating under 26 feet of head, Boyden's turbine generated 55 horsepower, Leffel's, 96 horsepower, and McCormick's, 266 horsepower. And it was these later improvements which brought the Francis turbine international fame and made it the most widely used design to the present day (82).

The millwrights' improvements in the Francis turbine have been variously ascribed to intuition and 'cut and try'. None of these men published their methods, but the case of James Leffel of Springfield, Ohio, is suggestive. Leffel introduced two critical improvements that have remained fundamental. Not only did he, like Swain, increase the turbine's capacity, but he introduced a second innovation, pivoted guide vanes (83). These greatly improved turbine efficiency at part gate, since they allowed the angle of entry to be varied as the speed of the wheel changed.

We have a clue to Leffel's methods. In the case of *Parker vs. Hulme* in 1849, Zebulon Parker introduced in evidence a glass-sided device which demonstrated, by means of small pellets immersed in the water, the fluid behavior involved in the operation of his reaction wheel (84). It is likely that he simply put a glass side or window on the miniature testing flume he had used as evidence in his earlier patent case, *Parker vs. Hatfield* in 1844. Parker's fellow Ohioan, James Leffel, had every reason to follow his patent litigation. Leffel had patented an unsuccessful reaction wheel, and he planned eventually to return to the manufacture of water motors. Thus, while Boyden missed the significance of Parker's innovation, Leffel apparenty did not. Leffel built a miniature, glass-sided testing flume and a Prony dynamometer with which he made more than one hundred experiments (85). These experiments probably go far to account for the spectacular success of Leffel's turbine, which outsold all of its rivals in America for a number of years. The direct observation of the complex flow of fluids has become one of the standard tools by which modern scientists study aerodynamic as well as hydraulic problems. In this, as in other matters, American millwrights were developing methods that can best be described as 'scientific', in order to generate the knowledge needed by technology.

Notes and References

Abbreviations used in References

ABFRC   U.S. Circuit Court, Pennsylvania, Eastern District, Equity Cases, 1790–1911, R. G. 21, Archives Branch, Federal Archives and Records Center, General Services Administration, Philadelphia, Pennsylvania

FARC   U.S. Circuit Court, Ohio, Southern District, Chancery Record, Series 1, Volume 7, FARC 924424R, Archives Section, Federal Records Center, General Services Administration, Chicago, Illinois.

1. Edwin Layton, 'Technology as Knowledge', *Technology and Culture* 15, 1974, pp. 31–41.
2. *ibid.*
3. Thomas P. Hughes, *Elmer Sperry; Inventor and Engineer*, Baltimore: Johns Hopkins University Press, 1971.
4. Edwin Layton, 'Mirror-Image Twins: The Communities of Science and Technology in 19th Century America', *Technology and Culture*, 12, 1971, pp. 562–580.
5. 'Parker vs. Brandt, *et. al.*', *Federal Cases*, vol. 18 (Case No. 10,727, April, 1850), p. 1117.
6. 'Report on Parker's Water Wheel', *Journal of the Franklin Institute* (3rd ser.), 12, 1846, pp. 35–37.
7. Arthur T. Safford and Edward Pierce Hamilton, 'The American Mixed-Flow Turbine and Its Setting', *Transactions of the American Society of Civil Engineers* 85, 1922, p. 1242.
8. Oliver Evans, *The Young Mill-Wright and Miller's Guide*, Philadelphia: Carey and Blanchard, 9th edn., 1836, pp. 159–160, fig. 30, plate IV.
9. Bernard Forest de Bélidore, *Architecture Hydraulique*, 4 vols, Paris: nouvelle édition, avec les notes et additions par M. Navier, Firmin Didot, 1819, I, book 2, ch. 1, plate 5.
10. Jean-Charles de Borda, 'Mémoire sur les Roues Hydrauliques', *Mémoires de l'Academie Royale des Sciences* (1767), pp. 270–287.
11. Eda Fowlks, 'The Invention of the Hydraulic Turbine', unpublished seminar paper, University of Minnesota, 1976.

12. Borda, 'Mémoire sur les Roues Hydrauliques', pp. 274–277, 283–284, and Charles C. Gillispie, *Lazare Carnot, Savant*, Princeton: Princeton University Press, 1971, p. 60.

13. Benoît Fourneyron, 'Mémoire sur l'application en grand, dans les usines et manufactures, des turbines hydrauliques ou roues à pallettes Courbes de Bélidore', *Bulletin de la Societe D'Encouragement Pour L'Industrie Nationale* **33**, 1834, pp. 4–17.

14. Louis Navier, notes and illustrations in Bélidore, *Architecture Hydraulique*, I, pp. 451–463, book 2, ch. I, figure 8, added plate D; Borda, 'Mémoire sur les Roues Hydrauliques', p. 285.

15. In 1828 when he began his experiments on the scroll, Zebulon Parker called the radial outward-flow wheel the 'common' reaction wheel. See Zebulon Parker, 'Sketch of the Invention of Parker's Water Wheel', *Journal of the Franklin Institute*, (3rd ser.), **22**, 1851, p. 48.

16. Fourneyron, 'Mémoire sur l'application en grand dans les usines et manufactures, des turbines hydrauliques', pp. 4–17.

17. Several of the depositions used in Parker's various patent suits were concerned with Tyler's 'Wry-Fly'. Elisha Hitchcock, testified that he had assisted Tyler in building what was evidently his first 'Wry-Fly' wheel in 1798. He gave a fairly complete description of this wheel, but the model he submitted in evidence has not been preserved. Hitchcock built a total of approximately 100 of these wheels. Hitchcock's description may be supplemented by that of James Sloan who made comparative tests of the 'Wry-Fly', Parker's wheel, and the 'common' reaction wheel in 1844 using a model testing flume with Proney Dynamometer. ('Ezekiel [sic] Hitchcock's Affidavit', June 4, 1850, *Parker vs. Sundry Defendants*, affidavits for cases 3–98, box 514, file 3, ABFRC, and 'Deposition of James Sloan', *Parker vs. Hatfield*, pp. 73–74, FARC.)

18. Parker, 'Sketch of the Invention of Parker's Water Wheel', p. 48.

19. The illustration is in 'Depositions of Witnesses. . . Parker vs. Hulme, . . . Oct. 27, 1849', *Parker vs. Brandt et al.*, box 512, file 32, ABFRC.

21. Evans, *The Young Mill-Wright*, pp. 9–152.

22. *ibid.*, pp. 78–79.

23. *ibid.*, pp. 122–152, *passim*.

24. 'Motion and Forces – Action and Reaction', undated ms (c. 1846), Committee on Science and the Arts, CSA 490, Franklin Institute Archives, Philadelphia, Pennsylvania.

25. Zebulon Parker to William Hamilton, Nov. 20, 1841. Committee on Science and the Arts, CSA 307, Franklin Institute Archives.

26. Zachariah Allen, *The Science of Mechanics, as Applied to the Present Improvements in the Useful Arts in Europe, and in the United States of America: Adapted as a Manual for Mechanics and Manufacturers and containing Tables and Calculations of Practical Utility*, Providence: Hutchens and Cory, 1829.

27. The nature of their testing apparatus is obscure. Austin Parker's original idea or invention apparently dates to the period when he and Zebulon were perfecting their scroll case, but it was probably not built. He hoped to get money from the Ohio legislature to perfect it and carry out systematic tests of water wheels. (See Austin Parker to Zebulon Parker, Aug. 21, 1829, Oct. 9, 1830. letters 'A', and 'B', *Parker vs. Hatfield*, pp. 26, 27, FARC.)

28. 'Exhibit A', *Parker vs Hatfield*, pp. 75–76, FARC.

29. 'Parker vs. Hulme', *Federal Cases*, vol. 18 (case no. 10,740, Nov., 1849), pp 1143–1144.

30. 'Deposition of Conington W. Searle', *Parker vs. Hatfield*, p. 55, FARC.
31. Austin Parker to Zebulon Parker, Aug. 28, 1831 letter 'D', *Parker vs. Hatfield*, p. 31, FARC.
32. 'Depositions of Witnesses . . . Parker vs. Hulme', p. 8. *Parker vs Brandt, et al.*, box 512, file 32, ABFRC.
33. *ibid.*, pp. 13–14.
34. Austin Parker to Zebulon Parker, May 4, 1831, letter 'C', *Parker vs. Hatfield*, p. 29, FARC.
35. Austin Parker to Zebulon Parker, Aug. 28, 1831, letter 'D', *Parker vs. Hatfield*, p. 30, FARC.
36. Austin Parker to Zebulon Parker, Aug. 21, 1829, letter 'A', *Parker vs. Hatfield*, p. 26, FARC.
37. U.S. Patent, no. 1,658, June 27, 1840; unnumbered patent, Oct. 19, 1829, copy in 'Bill in Equity', *Parker vs. Brandt et al.*, box 512, file 39, ABFRC.
38. Austin Parker to Zebulon Parker, May 4, 1831, letter 'C', *Parker vs. Hatfield*, p. 29, FARC.
39. Austin Parker to Zebulon Parker, Aug. 28, 1831, letter 'D', *Parker vs. Hatfield*, p. 31, FARC.
40. Austin Parker to Zebulon Parker, Aug. 21, 1829, Aug. 17, 1833, letters 'A' and 'G', *Parker vs. Hatfield*, pp. 23, 37–39, FARC.
41. Ellwood Morris, 'Remarks on Reaction Water Wheels Used in the United States and on the Turbine of M. Fourneyron', 3rd ser., vol. 4 (Oct., 1842), pp. 224–225.
42. 'Report on Parker's Water Wheel', *Journal of the Franklin Institute* (3rd ser.), **12**, 1846, pp. 35–37. Most early tests of turbines using the Proney Dynamometer gave efficiencies that were too high. Zebulon Parker got 75 percent efficiency in his own tests of his reaction wheel, and he was upset by the low efficiencies in the Franklin Institute's tests. He found that the scroll cases had been incorrectly constructed, producing extensive eddying or turbulence on entrance to these wheels. But the figures obtained were probably about correct, the two sorts of errors canceling out, approximately. In the case of *Parker vs. Hatfield*, Parker had tests performed by another millwright, James Sloan, and introduced into evidence. The efficiencies measured for the common reaction wheel, Tyler's 'Wry-Fly', and Parker's wheel were 50.3, 57.5, and 74 percent, respectively. Assuming that the tests were internally consistent, a reduction by 1/7, would give the more probable results, 37.2, 49.2, and 64 percent, respectively, for these three wheels. (See Zebulon Parker to John Cresson, Dec. 22, 1845, Committee on Science and the Arts, CSA 444, and 'Deposition of James Sloan', and Exhibits 'A' and 'B' in *Parker vs. Hatfield*, pp. 73–77, FARC.)
43. 'Hydraulics', *Scientific American* (n.s.) **6**, 1851, p. 256.
44. Zebulon Parker to William Hamilton, May 24, 1841, and broadside 'Parker's Patent, Percussion and Reaction Wheels', Committee on Science and the Arts, CSA 307, Franklin Institute Archives.
45. Austin Parker to Zebulon Parker, Aug. 21, 1829, letter 'A', *Parker vs. Hatfield*, p. 23, FARC.
46. 'Hydraulics', *Scientific American* (n.s.) **6**, 1851, p. 256.
47. Parker, 'Sketch of the Invention of Parker's Water Wheel', p. 49.
48. 'Hydraulics', *Scientific American* (n.s.) **6**, 1851, p. 256.
49. *ibid.*

50. Zebulon Parker to William Hamilton, May 24, 1841, Committee on Science and the Arts, CSA 307, Franklin Institute Archives.

51. Dennis G. Shepherd, *Principles of Turbomachinery*, New York: Macmillan, 1956, p. 53.

52. Zebulon Parker to William Hamilton, May 24, 1841, Committee on Science and the Arts, CSA 307, Franklin Institute Archives.

53. For a brief overview of the development of Lowell, see Robert F. Dalzell, Jr., 'The Rise of the Waltham-Lowell System and Some Thoughts on the Political Economy of Modernization in Ante-Bellum Massachusetts', *Perspectives in American History* **IX**, 1975, pp. 229–268.

54. James B. Francis to Washington Hunt, Jan. 20, 1858, vol. DA-5, Proprietors of Locks and Canals Papers, Baker Library, Harvard University.

55. James B. Francis, *Lowell Hydraulic Experiments*, (2nd. edn.), New York: Van Nostrand, 1868, pp. 1–2.

56. *ibid.*, p. 2.

57. Elwood Morris, 'Experiments on the Useful Effect of Turbines in the United States', *Journal of the Franklin Institute* (3rd ser.), **6**, 1843, pp. 377–79.

58. 'Article on Turbines Written for the American Cabinet', unpublished ms, Papers of Uriah Atherton Boyden, National Museum of History and Technology, Smithsonian Institution, Washington, D.C.

59. James B. Francis, 'Address', *Transaction of the American Society of Civil Engineers* **10**, 1881, p. 192.

60. James B. Francis to H. M. Birkinbine, Dec. 18, 1858, vol. DA-5, Proprietors of Locks and Canals Papers.

61. Francis, *Lowell Hydraulic Experiments*, p. 52.

62. James B. Francis to Luigi D'Auria, Feb. 25, 1878, vol. A-16, Proprietors of Locks and Canals Papers.

63. 'Articles on Turbines Written for the American Cabinet', p. 5.

64. Boyden to Joseph Henry, March 20, 1856, file: 'Copies of Letters Written, 1850–1859', Boyden Papers.

65. Boyden to Francis, Aug. 14, 1867, file: 'Letters Sent, 1860–1869', Boyden Papers.

66. I. P. Church, 'The Alleged "Remarkable Error in the Theory of the Turbine Water Wheel"', *Journal of the Franklin Institute* **117**, 1884, p. 333.

67. Shepherd, *Principles of Turbomachinery*, pp. 35–39.

68. Francis, *Lowell Hydraulic Experiments*, pp. 44–52.

69. *ibid.*, plate VI, fig. 1.

70. U.S. Patent number 5,063, April 17, 1847.

71. U.S. Patent number 10,026, Sept. 20, 1853.

72. U.S. Patent number 5,090, May 1, 1847.

73. Francis, *Lowell Hydraulic Experiments*, pp. 3–5, 209–231.

74. 'Hydraulics', *Scientific American* (ser. 2) **6**, 1851, p. 280.

75. 'Exhibit A', *Parker vs. Hatfield*, pp. 75–76, FARC.

76. Francis, *Lowell Hydraulic Experiments*, pp. 71–145.

77. 'Report on Parker's Water Wheel', p. 37.

78. Sub-Committee report and drawings, Committee on Science and the Arts, CSA 572, Franklin Institute Archives.

79. Francis. *Lowell Hydraulic Experiments*, p. 55.

80. *ibid.*, table VI, pp. 66–67. The maximum efficiency was about 79 percent.
81. The work of Swain and McCormick is summarized in John R. Freeman, 'General Review of Current Practice in Water Power Production in America', *The Transactions of the First World Power Conference*, 5 vols., London: Percy Lund Humphries, n.d., II, pp. 379, 380.
82. Samuel Webber, 'Water Power – Its Conservation and transmission', *Transactions of the American Society of Mechanical Engineers* 17,1895–1896, p. 49.
83. Stafford and Hamilton, 'The American Mixed-Flow Turbine and its Setting', pp. 1261, 1308, 1319.
84. 'Parker vs. Hulme', *Federal Cases*, vol. 18 (case no. 10,740, Nov., 1849), pp. 1143–1144. It is possible that Parker borrowed this idea. In 1851 *Scientific American* reported that Haviland and Tuttle had displayed their wheel "two years ago in New York", and that they had mounted a "small working model in a glass case, whereby all its motions and behavior could be distinctly observed". (See 'Hydraulics', *Scientific American (n.s.)* 6, March 8, 1851, p. 200.)
85. W. W. Tyler, 'The Evolution of the American Type of Water Wheel', *Journal of the Western Society of Engineers* 3, 1898, p. 890.

SCIENCE, TECHNOLOGY AND ECONOMICS: THE INVENTION OF RADIO AS A CASE STUDY

HUGH G. J. AITKEN

Amherst College

Economic history is one of the hybrid disciplines, conceived in the late nineteenth century, offspring of a transitory union between two otherwise virtuous academic parents. It has, in its later career, regularly jeopardized its slender claims to respectability by a reluctance to honor conventional boundaries. Economic historians are most at home in the disputed areas between disciplines. They cultivate the borderlands, the uncertain margins between history and economics, between the humanities and the social sciences.

With one part of his professional *persona*, the economic historian must be involved in the problem that is at the heart of modern economic analysis: how scarce resources are allocated among competing alternative uses. But he is also a species of historian; he must therefore also be interested in how these patterns of resource allocation have changed over time. Difficulties arise when these allegiances pull in different directions: when the concern with history conflicts with the paradigms of economic theory.

This conflict becomes particularly apparent when economic historians concern themselves with things that economic theorists take to be 'given'. These are the parameters that, in conventional theory, are impounded in the closet labelled *ceteris paribus*. Among them are changes in 'preferences' and changes in the 'state of the arts'. By changing preferences we mean not only the whims and vagaries of taste but also changes in the ends that economic activity serves — the priorities that are reflected in national policy. Economic theory prefers to assume that these preferences are 'given'. The economic historian, however, cannot avert his gaze so easily. He wants to know how these preferences took the form they did.

Krohn/Layton/Weingart (eds.), The Dynamics of Science and Technology.
Sociology of the Sciences, Volume II, 1978. 89–111. All Rights Reserved.
Copyright © 1978 by D. Reidel Publishing Company, Dordrecht, Holland.

It is the same with changes in the state of the arts – technological change in the broadest sense. Economic theory proceeds on the assumption that production functions are known, so that choices as to how scarce factors of production are to be combined can be made by reference to relative prices and marginal productivities. But there is no economic theory of how production functions change, of how new production functions are brought into existence. This set of problems is left to the economic historian.

Economic historians who take up the analysis of technological change face two major problems. The first is the lack of a theoretical model. There are clues in the work of Schumpeter, Usher, Schmookler, Rosenberg, and a few others (1). Some good work has been done on technological diffusion, which lends itself to mathematical treatment. But there is little theoretical guidance available on how technological change originates. What determines the kinds of technological change that happen? What creates the array of production possibilities that theory takes as given?

The second problem is usually inadequate competence in the particular field of technology involved. The training through which economic historians pass *en route* to professional certification seldom includes exposure to industrial technology. Private study can fill some gaps but cannot provide the intimate 'hands on' knowledge of machines and processes that is required. Training in the history of science is also usually inadequate.

This combination of difficulties is enough to deter the bravest. It is easy to make a fool of yourself when you stray 'off the reservation'. The subject matter is difficult; theory furnishes little aid; and one's training is usually inadequate. Economic historians are accustomed to having economists and historians looking over their shoulder, the one insisting on rigor and quantification, the other forever harping on interdependence and the 'seamless web' of history. But when the economic historian tackles the history of technology there are new arbiters to be satisfied.

My own interest is in the history of radio technology. This interest has origins partly personal and partly professional. The personal origins are unlikely to interest the reader, except to the extent that they led to the acquisition of some degree of skill in radiocommunications and thus helped to counteract the limitations on technical competence referred to above. Professionally, I have been concerned with how economic historians should tackle the analysis of technological change, bringing to the task their own intellectual orientation and expertise. This is, I believe, one of the major challenges facing economic history today; the response it evokes will be important not only to that discipline but also to historians of technology generally.

My purpose in the present article is two-fold: first, to sketch briefly the key events and processes in the development of radio technology up to the time of its commercial acceptance – that is, roughly to 1900; and second, to raise some questions about the relations between science, technology and economic life in that period. The empirical material may already be familiar to some; readers who wish a more detailed account are referred to the author's *Syntony and Spark* (New York, 1976). The more analytical sections may eventually contribute to the construction of a general model of technological change in terms of information theory.

*   *   *   *

Who invented radio, and when this happened, are questions that have no simple answers. If we mean the basic theory of the electromagnetic field, credit goes to James Clerk Maxwell and the appropriate date is 1864, when Maxwell read his paper, 'A Dynamical Theory of the Electromagnetic Field', before the Royal Society. If we mean the creation of a feasible technology for transmitting and detecting electromagnetic radiation at known frequencies, we move forward in time to 1888 and we refer to Heinrich Hertz and his work at Karlsruhe. If we mean the use of this technology to transmit messages, the evidence points to Oliver Lodge and the demonstrations he gave in London and Oxford in the summer of 1894. If we have reference to proof of commercial feasibility in competition with alternative modes of communications, the man to be recognized is Guglielmo Marconi and 1897, when Marconi's Wireless Telegraph and Signal Company was incorporated, is a defensible date. And finally, if we mean by radio not point-to-point communication but radio broadcasting (which is probably what most people understand by the term), we move all the way up to 1920 and credit goes to Frank Conrad, the Westinghouse Company, and station KDKA in Pittsburgh (2).

The point of this exercise is not to demonstrate the arbitrariness of dating and assigning personal credit for an invention but rather to emphasize that we are dealing not with an event but with a process – a process in the course of which the thing invented changed, coming in time to take forms and serve functions that were no part of the original expectations. We can interpret this process in terms of exchanges of information. The 'invention' can be seen as the progressive emergence of new configurations of knowledge – in science, in technology, and in economic life.

Hertz's purpose in conducting his experiments was a purely theoretical one. There was no thought of 'practical applications'. What Hertz intended to do was to test Maxwell's model of the electromagnetic field, at that time no more than an intellectual construct. 'Testing' in this context meant, firstly, translating the model into operational form, so that measurable relationships between key variables could be specified; and secondly, devising apparatus that would permit these measurements to be taken. Three related hypotheses were involved: that electromagnetic fields could be generated by the acceleration and deceleration of electric currents; that these fields could be propagated either along a conductor or through a dielectric; and that their velocity of propagation was finite – specifically, that it was the same as the speed of light. Hertz realized that all three hypotheses could be tested simultaneously, and the possibility of decisive disproof of Maxwell's model realized, if he could demonstrate the generation and presence of electromagnetic waves in free space and measure their velocity. His clarity of vision on this point was what set him apart from contemporaries such as Lodge (3).

The design and results of Hertz's experiments have been described many times, and this is not an appropriate place for extended comment on them. One point, however, must be emphasized: clarity of insight into the theoretical requirements for validation or rejection of Maxwell's field theory was not in itself sufficient. Also required was the creation of an experimental technology – a technology that would generate the phenomena called for by the model in a predictable and controllable way, detect them, and enable them to be measured with precision. Hertz was by no means the first individual to generate electromagnetic waves; but he was the first to do so in such a way that a particular model of the physical universe could be put to the test, and that depended on his specific laboratory technology. Hertz's experiments might have led to a decisive rejection of Maxwell's model; that was their importance for science. The feasibility of the experiments depended on technological creativity. The technology involved, however, was technology of a particular kind: the technology of observation and measurement that was and is an intrinsic part of scientific research. Hertz's discoveries were contingent upon a merging of scientific and technological knowledge in this sense.

Hertz's work, in short, demonstrated a two-way transfer of information between science and technology. Whether Hertz realized it or not, he had created in embryonic form a means of communicating through space without wires. He had, in one sense of the phrase, 'invented' radio, in that a construction of pure thought – a classic example of the outputs of pure science – had been translated into usable technology, a particular *gestalt* of hardware. This hardware could be used, and would be, for purposes never contemplated by the scientists who had devised and elaborated the model. Once part of the informational capital stock of the technological system, it became available for combination with other bits of information – in this case, information about problems of communicating over distance. The forward transfer had moved information out of one context into another. Simultaneously, however, there had occurred a reverse transfer, for Hertz had made available to the world of science a technology it had not possessed before, a technology by which electromagnetic radiation at known frequencies could be generated, transmitted, detected, and measured. This was, by any reasonable standard, a major increment to the informational stock of science. *This* part of Hertz's work was not a scientific achievement in the strict sense but rather a technological one; nevertheless, it was a major piece of new information for science to use, and the explosion of intense research activity on radiation that promptly followed is testimony to its scientific importance.

Hertz was not, of course, the only physicist working on these problems in the late 1880s. Oliver Lodge in particular had at an early age selected as his primary research objective the testing of Clerk Maxwell's model, and after his appointment to the University College of Liverpool in 1891 he gave it as much time as his teaching and other distractions permitted. His approach differed significantly from that of Hertz, of whose work he knew nothing at the time. Both men used an induction coil and a spark gap to generate electromagnetic disturbances but while Hertz connected this apparatus to a dipole antenna and radiated waves into space, Lodge connected the spark gap to a pair of long wires and used these conductors as wave-guides. Theoretically the difference in technique was unimportant. In terms of its later consequences, however, Lodge's choice of technique was significant. In the first place it brought into prominence the phenomena of tuning, or what Lodge called syntony. By varying the length of the wires Lodge could create tuned circuits that resonated at the frequency of the spark discharge; and by coupling two such circuits together, he could excite a spark in one when, and only when, it was tuned to resonance with the other. This phenomenon was to prove pivotal to later use and allocation of the radiofrequency spectrum. Secondly, Lodge's experimental technique placed him squarely in the world of telegraphic communication. His original intention had been to create a spark at one end of a long telegraph circuit and measure how long it took for the disturbance to reach the other end. The long wires he used in his laboratory, left open at the far end so that the disturbance was reflected back to the source, were essentially substitutes for this long telegraph circuit (4).

Hertz's method involved radiating electromagnetic disturbances from an antenna through space. Lodge's involved transmitting them along wires. There is no question which called for the greater creative imagination. Compared to Hertz's, Lodge's approach was pedestrian. Nor is there any question which laid the basis for future technological change. Hertz's technique looked forward to the technology of radio, Lodge's back to the technology of the wired telegraph. But, for our purposes, these are not the central points. Precisely because Hertz's method involved radiation through space, its connection with communications, with signalling over distances, was not immediately or readily apparent: such a medium had never been used for that purpose before, and to see that it could be so used called for a second leap of creative imagination. Lodge, in contrast, was working with the hardware of a familiar technology. No one could inspect his apparatus, or even see it illustrated in a published report, without having its possible application to communications brought forcibly to mind.

This is not to say that Lodge himself showed initially any interest what-

ever in commercial development. Communications technology as such
was not one of his original concerns. Lodge did have, however, throughout
his life, a great interest in public lectures and demonstrations, and
considerable skill in presenting them, and it was in the course of preparing
for one such demonstration that the first application of his apparatus to
signalling took place. This was in 1894. By that date Lodge had made
refinements in Hertz's apparatus, in particular by using the coherer as a
detector and by introducing sharper tuning of the radiating and receiving
circuits. These were, it should be emphasized, technological advances;
their effect was to convert laboratory equipment designed for a different
purpose into a communications system that, though primitive, worked.

This apparatus was first shown in the course of a public lecture at the
Royal Institution in June, 1894. Later that month, at a 'ladies' evening' at
the Royal Society, a small portable receiver was demonstrated. With this
receiver and the usual spark gap transmitter Lodge claimed that ranges of
half a mile had been achieved. At the Oxford meeting of the British
Association in August of that year Lodge publicly demonstrated transmission and reception of Morse code signals over approximately the same
distance. The lectures that accompanied these demonstrations dealt with
the electromagnetic theory of light – still, for a scientist, the central issue –
but what was demonstrated was a system for the transmission and
reception of information by radio waves.

It is interesting to observe the growing role in these demonstrations of
standard telegraph apparatus and methods — the signalling key, a mirror
galvanometer originally designed for the Atlantic cable, and the Morse
code itself. What was being shown was a melding of the new Hertzian
apparatus with the wired telegraph. In abstract terms, inputs of information from the Faraday-Maxwell-Hertz tradition in science were being
absorbed into the information stocks of communications technology. And
as this occurred the 'steering' effect of information from the economic
system first became evident – muted, to be sure, and little heeded at first
but nonetheless present. The man responsible was a certain Alexander
Muirhead, partner in a manufacturing firm that produced instruments for
use in cable telegraphy. Muirhead apparently furnished Lodge with the
telegraphic equipment needed for his demonstrations and, about the time
of the Oxford lecture, suggested that he and Lodge should form a

syndicate for the manufacture and sale of 'wireless' apparatus. Lodge was unresponsive at the time but within a few years, after Marconi's arrival in England, the Lodge-Muirhead Syndicate was to become a reality, its principal assets being Lodge's patents on 'syntonic' telegraphy. Patents, which Lodge had previously spurned as a distraction from the true calling of a scientist, were now necessary if property rights were to be defended.

Commenting several years later on the process by which Marconi, an unknown foreigner, had so quickly achieved a dominant position in British radiotelegraphy, Lodge gave two reasons to explain why he himself had not recognized the opportunity. First, he said, he had not thought it proper to involve himself in the details of practical applications. He was a scientist; if there was a need for new signalling apparatus, let it be attended to by the appropriate public officials, specifically by the British Post Office, which had a statutory monopoly of wired telegraph systems. But secondly, it had not been at all clear that any such need existed. What was the point of investing time and money in developing 'wireless' telegraphy when the existing wire systems worked so well? ". . . so far as the present author was concerned he did not realize that there would be any particular advantage in thus with difficulty telegraphing across space instead of with ease by the highly developed and simple telegraphic and telephonic methods rendered possible by the use of a connecting wire" (5).

We assume too readily, when a new technology becomes available, that it has appeared in response to some kind of social need, and that the functions which, in the course of time, it comes to serve are those which originally called it into existence. Neither of these relations necessarily holds, and the early history of radiocommunications provides one clear instance to the contrary. The relevant technology emerged as an un-anticipated by-product of scientific research undertaken with a quite different end in view. And the uses which it originally served were ones which before long were to seem relatively trivial. Even these uses were not apparent to all. To find an economic place for the new technology called for imagination and salesmanship. It called for the kind of entre-preneurship that could create needs, not merely serve them. It called for the merging of the information contained in the new technology with the information provided by the economic system. This merging did not happen automatically; it was the result of creative insight, the ability to arrange information into new patterns, and of a determined effort of will, the ability to make events follow a path already settled in the mind.

Guglielmo Marconi played this role. Introduced by his tutors to the work of Heinrich Hertz, Marconi immediately thought of communications – of communications over long distances, without interconnecting wires. To him, the information he got from Hertz *meant* a communications system. His several biographers and his own account tell vividly of his amazement that what was evident to him was not so to others, and of his anxiety lest someone better situated or better equipped than he was would anticipate him in showing what Hertzian waves could do. He was clear, too, about what was needed to make commercial development possible. He needed to achieve greater range – much greater distances than Hertz had needed in his laboratory or than Lodge was toying with at Liverpool and Oxford. And he needed to find a market, which is to say that he needed to identify communications needs that the existing technology of wired telegraphy was serving poorly or not at all. These requirements led him to higher power levels, new forms of antennas, and more sensitive detectors. They led him, in short, to refine the technology in directions indicated by prospective economic use. The steering effect of the market, to which Lodge was insensitive, was from the beginning the dominant influence on Marconi's work (6).

The progressive development of Marconi's equipment has been analyzed in detail elsewhere and need not detain us here. The question of the markets to be served is more directly relevant. What Marconi had to locate were the 'holes' in the existing communications technology – functional areas where service was inadequate or non-existent. Over land, wired telegraphy and telephony systems generally gave good service, but the situation was different with maritime communications. Over very long distances, and where the volume of traffic could justify the investment, submarine cables were feasible. But they did not meet the need for communications with and between ships at sea; and in certain other cases, such as communications with offshore islands and lighthouses, a communications system that dispensed with connecting wires was an intriguing possibility. In particular, there was a large potential market for facilities that offered faster, more reliable, and more complete marine intelligence – information on ship movements and on emergencies at sea. And there was a market for naval communications of immense potential. The strategic deployment and tactical manoeuvering of naval squadrons were hardly issues that, in the closing decades of the nineteenth century, any of the world's major powers could ignore.

Considerations such as these took Marconi, after an initial offer to the Italian government, to England: involved him promptly with the British Post Office and the Admiralty; and led him to the issuance of his first British patents – all-inclusive documents in which he claimed property rights to almost every advance made in radio technology up to that date, except tuned or syntonic circuits. It is remarkable, indeed, how quickly and effectively Marconi won recognition. He had, it is true, important allies, notably William Preece, chief engineer of the Post Office, and Captain Henry Jackson of the Royal Navy, who had been working independently toward the use of radio at sea. But the support he received from men such as these suggests that he had analyzed accurately the economic conjuncture into which the new technology had to be fitted. Preece, concerned at this time with the communications problems of off-shore islands and lighthouses, had been experimenting with inductive telegraphy but with unsatisfactory results. Jackson had been having trouble getting the ranges necessary. To these men Marconi's arrival was timely and helpful. He had a working system, one reliable enough to survive field testing. He was anxious, at least initially, to work through official channels. And he knew what his equipment could and could not do.

What was happening was the successful transfer of new technology into the economic system. Simultaneously that technology was undergoing refinement in terms of the economic functions it was to serve. Marconi played a central role in both respects. In the long run some of the decisions he took were questionable. In particular, Marconi may have moved too quickly out of the high frequency and very high frequency areas of the radiofrequency spectrum – frequencies which had much to commend them for naval and commercial use – in the belief that longer wavelengths meant longer distances. And it may well be that, after his negotiations with the Post Office broke down, he led his Company too quickly into a competition with the submarine cables for trans-Atlantic traffic for which it was technologically not prepared. But these ventures, including Marconi's obsessive search for greater distance, paid important technological and scientific dividends, notably the discovery of ionospheric propagation, as well as publicizing very effectively the potentials of the new medium. The solid foundation of the fortunes of the Marconi system remained, in any event, its dominant position in commercial marine traffic. Seen from this point of view, the turning point was probably the signing of an exclusive traffic-handling contract with Lloyds of London in 1901 – an event which, both symbolically and in fact, marked successful transfer of the new technology into economic use.

From this point until the introduction of the vacuum tube in 1906 the technology of radio experienced no fundamental change. There were important refinements: improvements in antenna design, more sensitive and reliable detectors, and an increasing emphasis on tuned circuits as the spectrum became crowded. But the basic technical limitation lay in the fact that there was no means of amplifying a radiofrequency signal. Greater and greater power at the transmitter, larger and larger antennas, a progressive move to lower and lower frequencies – these were all means of compensating for the lack of a radiofrequency amplifier. Likewise, as long as the spark discharge remained the standard means of generating radio waves, there could be no transmission of voice or music, for the wave train so generated was in reality a chain of rapidly decaying pulses which it was impossible to modulate. Telegraphy, therefore – the keying of the transmitter on and off – remained the sole method of radio communication. Attempts to escape from this impasse included the arc transmitter and the radiofrequency alternator – transitional devices which made possible higher power and a closer approximation to a continuous wave. But the decisive breakout came only with the vacuum tube. As a diode detector it was more sensitive than any of its predecessors. But more important, with a third element or grid added, it became an amplifier and with sufficient gain and positive feedback it became an oscillator, capable of generating a continuous sine wave at radio frequencies. This was, it should be added, a distinctively technological achievement; it called for no advance in pure science.

With the triode vacuum tube radio technology entered its second phase, a phase from which it has emerged only recently with the invention of the transistor and the integrated circuit. Radiofrequency amplification and sustained continuous wave oscillations now became possible, bringing into the realm of technological feasibility the transmission of voice and music, low cost receivers in the home, and eventually radio and television broadcasting. By this time radio had become indispensable not only to mass entertainment and merchandising but also to the safety of ships at sea, the gathering and dissemination of news, and the conduct and prevention of war. Radio technology developed rapidly in response to these demands, reinforced at strategic junctures by advances in applied science such as those that made possible the heterodyne principle, frequency modulation, and television. In this stage of development inter-action between technology and the market was intense and continuous. Its importance was well recognized by those involved. Large sums were spent on research and development facilities. And on the investment decisions that determined which technological advances would be selected for commercial use and which would not, the fate of large corporations and of individuals depended. Whereas, in the initial phase, development had depended on mere trickles of information passing across the interfaces between science, technology, and the economy, now there was a flood so great that scanning it and filtering it for possible use became specialized functions, essential both for corporate survival and for effective planning of research. Nor was pure science isolated from these exchanges. As a recipient, science profited from the multitude of new devices that twentieth-century electronic technology generated. As an exporter of information, science provided the physical and chemical knowledge essential for the development of high vacuum tubes and eventually solid state devices such as the transistor.

The foundations for this phase of growth had been laid in the closing decades of the nineteenth century and in a much less highly institutionalized context. The interchanges of information that made possible the achievements of men like Hertz, Lodge, and Marconi seem, in comparison with what came later, narrow, sporadic, almost haphazard in their timing and incidence. One is struck by their personal and almost accidental character. Nevertheless, it is hard to deny their effectiveness. Diffusion of the new information was rapid. Only eight years elapsed between publication of Hertz's findings and the issuing of Marconi's first patent (7). Furthermore, in all respects but one, the appearance of randomness – the accidents of time, place and character – may be illusory.

We have concentrated here on Lodge and Marconi as the key agents in the transfer of information, the 'translators' who moved information from one domain to another. But other actors could have occupied the center of the stage: Popov, Branly, Braun, Arco, Ducretet, Shoemaker, de Forest, and Henry Jackson, to name only a few. If Lodge and Marconi had been absent, alternative transfer agents were available.

That Hertz's apparatus, by one channel or another, would promptly have been translated in a technology of communications seems indisputable. That a commercial use for this technology would promptly have been found in the absence of Marconi is more open to question. There were, after all, substantial vested interests to be reckoned with. Large amounts of capital had been invested in wired telegraph systems. And personal reputations had been staked on alternative approaches to 'signalling without wires'. It is impossible, in fact, to analyze the introduction of radio technology into practical use without being impressed by the talents of Marconi. These talents included not only a high degree of native engineering skill and a determination amounting almost to mulishness but also a keen sense of market opportunities and an ability to disarm opposition that defies easy explanation. When one considers the uncertainties present and the formidable vested interests to be dealt with, it seems beyond argument that successful translation of the new technology of radio into commercial use called for entrepreneurship of a high order. In any period of history such talents, involving as they do the ability to function effectively at the boundaries between highly specialized systems of human action, are the scarcest of scarce resources.

*   *   *   *

On the basis of this case study, what generalizations suggest themselves about technological change in the closing decades of the nineteenth century? Clearly one case cannot prove anything; equally clearly, our data relate to a specific historical context, and relationships evident in this case may not hold true for different times and places. Nevertheless, if analyzed with care and prudence, a particular case can be suggestive: it can tell us what to 'keep an eye on'; it can indicate which ways of looking at the data are likely to be helpful and which less so.

Underlying our narrative of the early history of radio technology has been an implicit model. This model sets up three systems of human action, which for convenience we have called science, technology, and the economy, and it states that these systems are related in specific ways. One of the ways – not the only one – in which the three systems are related is by flows of information. We have tried to make the course of technological change understandable – we have 'explained' it – by making statements about the direction, content, and timing of these flows. The test of the usefulness of the model is whether it enables us to see a meaningful pattern in events that otherwise would remain discrete, particular, and episodic.

What kind of 'systems' are these? From one point of view they are structures of social action, as Talcott Parsons for example conceived of them: functional sub-systems of the total society which set norms of behavior, rewarding certain modes of conduct and penalizing others (8). It is within these systems that the human actors in the drama of discovery, invention and innovation play their roles. From this point of view the three systems are differentiated from each other by their functions, by their value-orientations, and by their sanction patterns – the ways in which they discourage deviant behavior. Approved behavior for a businessman, for example, is unlikely to be in all respects approved behavior for a scientist. The scientific community assigns status and power by criteria which differ from those of the business community. And so on. These differences, of course, are historically and culturally variable.

A second way of looking at these three systems, and one that may be helpful to the historian of technological change, is to think of them as information processing systems. Each of them stores information: the stock of scientific knowledge, of technological lore, of business practice. Each of them generates information: additions to scientific knowledge, whether in the form of theories or of data; additions to the array of feasible technological devices and processes; and, in the case of the economy, additions to knowledge of production functions, preferences, and relative costs. Lastly, each system absorbs information: the economic system absorbs information from technology in the form of new products and processes; the technological system absorbs information from the economy in the form of data as to what technological outputs 'make sense' in terms of current costs and prices; and science, though perhaps the most self-sufficient of the three systems in terms of information requirements, nevertheless absorbs information – from technology in particular – in the form of empirical data, reports of anomalies, and knowledge of new devices and instruments that science can use.

The three systems, in short, are to be thought of as linked together by a network of information exchanges. These exchanges, as we shall later stress, are not limited to the forward transfers out of science into technology and thence into economic use that underlie most naive theories of technological change. Reverse flows are of vital importance and of intense interest to the historian. Furthermore, information is not the only commodity traded among the systems. Science and technology each receive their 'sustenance' from the economic system, either directly by the sale of outputs or indirectly via governmental transfers. And there are important transfers of personnel.

Discoveries, inventions, and innovations are the three terms normally used to characterize the emergence of novelty in science, technology, and the economy respectively. At the most abstract level of analysis, novelty emerges in each case when bits of information are integrated into new configurations. One way in which this may happen is when new items of information, imported from an outside system, are merged into the existing stock – when a new scientific discovery, for example, makes possible a novel configuration of technology. In principle there is no reason why the elements of new information that trigger the creative break-through should always be the results of imports from outside the system. There have been many technological inventions that required no new inputs of information from science; and there have been many economic innovations which required no importation of new knowledge from technology. Each system is custodian of its own stock of information, and novelty may emerge from recombinations of bits of information already in that stock as well as from the integration of new information arriving from an external source. In certain cases, however, particular interest attaches to the catalytic effect of new information moving between the three systems, and to the human agents and institutions that effect the transfers. The early history of radio technology, which furnishes the historical focus of the present study, provides such a case.

An elementary knowledge of cybernetic theory would suggest that reverse or feedback flows of information are likely to be of some importance in analyzing how our three systems 'track', or in other words in explaining their trajectories through historic time. The interface between technology and the economy illustrates the process very clearly. Information moves in both directions across this interface. Technology exports to the economic system information on new products and processes; it imports information on prices, costs, and expected profitability. The interface, in short, serves as a screening and filtering locus. Not all new information generated by technology has economic use; not all information generated in the marketplace has technological relevance. Flows in both directions across the interface have to be filtered. If the filtering process is functioning effectively (and this is clearly a historical variable) relevant information gets separated from the accompanying noise; and the information that passes through the filter serves as 'tracking data' for the receiving system.

All interfaces between functionally specialized systems are characterized by problems of coding and decoding – which is to say, by problems of mutual intelligibility. As a system develops in scale, specialization and complexity, it develops its own language, convenient and easily comprehended by participants, obscure and ambiguous to outsiders. And it develops a characteristic structure and sub-culture, one of the components of which is a specialized language. Communication among the three systems requires therefore the translation of information out of one language and into another.

Such problems of translation inevitably occur at the interface between technology and the economy, and variations in the way these problems are handled may have much to do with differences between countries and between industries in entrepreneurial receptivity to technological innovation. Between these two social sub-systems, however, there are structural correspondences that serve to ease difficulties of communication. Both, for example, use property rights to assign scarce resources among alternative uses, and as a legitimate source of income for participants. Both are highly pragmatic in their orientation. And both are organized to serve the kind of social need that can be reflected in prices and profit opportunities. Technology is, in short, Janus-faced: it presents one image to the world of science and another to the world of business. Each of these images duplicates certain of the features of the system technology deals with, and these similarities help to ease mutual interaction.

We are accustomed to thinking of technology as responsive to signals from the economic system; and in the contemporary world, where governments play a large role as purchasers of new technology, this feedback relationship has become very evident. Over much of human history, however, this was not so; in historical perspective, the close linkage between technological change and economic 'need' is a relatively recent phenomenon, and one which goes a long way toward explaining the sustained rates of economic growth in the industrialized nations over the last two centuries. We have learned, it would appear, how to organize the two-way transfers of information at the interface between technology and the economy.

Reverse transfers of information from the economy into science are harder to grapple with. Some highly idealized models of 'pure' science, indeed, stipulate that few if any such transfers take place. Science, we are sometimes told, is virtually self-sufficient in its information requirements, in the sense that the only data scientists 'hear' are data generated by scientists themselves under conditions of controlled experiment and observation. Furthermore, science steers its own course. The problems science chooses to tackle are problems set by science itself – indeed, are often problems only *to* science. Only in exceptional circumstances does information generated 'outside' science influence the tracking of scientific activity – some extreme national emergency, perhaps, or the persistence of anomalous results that conflict with orthodox scientific principles. Further, it has been held that science, unlike technology, is insulated from economic incentives and pressures by the fact that its outputs are not for sale for a price; new scientific knowledge cannot be property in the sense in which a patentable discovery in technology can. The value system of science and its institutional set-up serve to screen it from economic pressures and to reinforce its self-steering character. And, paradoxically, while the high productivity of technology is often attributed to its close links to the marketplace, the high productivity of modern science is often ascribed to its insulation from the market.

The danger in this mode of analysis lies is attributing universality to a situation that was historically and culturally specific. There was indeed a period in the history of western civilization – and our case study exemplifies it – when many scientists pursued their inquiries unconcerned (as far as we can tell) with whether their work had any commercial applicability at all. There was a time when a scientist thought it somehow disreputable, or at least a distraction from his true calling, to bother himself with patents and property rights. But these characteristics – and they could be elaborated – were specific to a particular historical period and locale. We are not entitled to elaborate them into a universal model of science in general. In particular, we should be wary of extrapolating this model to a period like the present, when the costs of scientific research have risen enormously and when only national governments and large corporations are capable of funding these costs (and carrying the risks that accompany them). In circumstances so changed, the 'steering' effect on science of information from the market (in which we must include governments) cannot safely be ignored. Most research in basic science today is related directly or indirectly to projects or 'missions' of one kind or another; most such projects are funded by governments, either directly or through the tax system; and the value-orientations and status-assigning processes of the scientific community appear to have adapted to this new context.

The seductive attractiveness of our case study is that it illustrates with such clarity the relations between science, technology and the economy in their simplest and most elementary form. And this simplicity can hold dangers for the historian. Let us look at some features of our case more closely. Clerk Maxwell's theory of the electromagnetic field is a classic example of the kind of output that pure science is supposed to produce: a conceptual scheme, a mathematical model whose empirical referents were unspecified, a construct the practical utility of which was completely irrelevant. With Hertz's experiments of 1888 the first step was taken in the process of translating this information 'forward' into the realm of technology. Hertz's creative achievement was, on the one hand, to translate Maxwell's model into a form in which it could be tested empirically and, on the other, to devise the system of scientific technology that made empirical testing possible. This achievement provided new information for science: the reverse flows were beginning to take place. But it also, in entirely typical fashion, provided new information for technology – information which, in combination with items of knowledge already in technology's inventory, made possible a novel method of communicating over distances through empty space. This had been no part of Hertz's intentions, nor of Maxwell's. The trajectory that their research was following was one which, in the domain of science, led ever farther into the theory of radiation. But the information they provided, translated into the technological domain, made possible a new trajectory of technological development: one which led eventually to what today we call radio.

Note that in this case there is not the slightest evidence that economic considerations influenced the research programs that Maxwell or Hertz were following. We are not saying that they lived and worked without economic support; we are saying that decisions as to 'what to tackle next' were taken by criteria other than profitability or the maximization of program budgets or anything of that sort. As far as we can tell, in this instance the insulation of scientific research from economic pressures and incentives was complete.

And this is precisely why Oliver Lodge is such a pivotal figure in our story. Lodge's status as a scientist was no less than that of Hertz. His qualifications were excellent. And in a number of respects – his reluctance to patent his discoveries, his aversion to litigation over property rights, his feeling of obligation to make what he knew immediately available to everyone – his personality characteristics typify the classical stereotype of the true scientist. But Lodge had also an intensely practical side to him. He thought science should be useful in solving problems – practical problems, not just theoretical ones. His first important research had been on electro-static precipitation, an industrial problem then as it is today. His work on electromagnetic radiation stemmed partly, it is true, from the same roots as Hertz's: the desire to test Maxwell's model. But it also stemmed from an immediate and urgent practical problem of his own day: how to improve protection against lightning strokes. Lodge's mind and temperament looked in both directions: toward pure science and toward technology.

It is, therefore, appropriate that it was Lodge who gave the first public demonstration of signalling by radio waves. That he did so almost casually in the course of a public lecture on the electromagnetic theory of light is also appropriate; and that he did not bestir himself to take out patents on his syntonic circuits – then as now a key element in radio design – until after Marconi burst on the scene, despite prodding from a business-minded associate, fits the same pattern. With Lodge, the forward transfer of information into the technological system becomes complete: we see a workable system of radiocommunications evolved and demonstrated; property rights are claimed and the steering impact of market signals becomes visible for the first time. But the pull of the market is still feeble. Signals from the economic system are incoherent, ambiguous, and full of noise – what is the new information good *for*? And the response of the key actor is vacillating. Conflicting value systems lead to delay and hesitation. What business does a true scientist have meddling in 'commercial applications'?

No such hesitations trouble Marconi. True, the market he aims for is not at first commercial message handling for profit but rather the needs of governments. But it is a market nonetheless. Marconi is a man with something to sell. Because of this sharpness of orientation the ambiguities that bother Lodge are nonexistent for Marconi, and the forward transfer from technology to the economic system is made quickly and with a sure touch. And the linkage remains firm thereafter. There is no technological development made by the Marconi enterprises from 1900 to 1918 that cannot be clearly related to commercial difficulties experienced or anticipated. Some of these technological advances generated significant feedback information for science – data on long-distance propagation via the 'Heaviside layer' is a good example – but they had their origin in problems of the marketplace.

The same patterns of interaction have remained strongly evident ever since, as the industry has matured, expanded and diversified. Ask for an example of a science-based industry today, and someone will surely mention radiocommunications or electronics. But the pyramid-like visual image conjured up by the term 'science-based' is in fact a caricature. The industry exists and functions as it does because science, technology and economics are, in its operations, woven together in a complex and mutually reinforcing web of interactions. This pattern was functional to the establishment of the industry and it has remained functional to its evolution ever since.

What general implications can we draw from this analysis? I suggest the following:

(1) The technological system can usefully be thought of as occupying a mediating or intermediary position between science and the economic system. In each of these systems information exists and is generated in a particular coded form. One of technology's functions is to translate new information generated by science into forms usable by the economy; and likewise to translate information generated by the economic system into such a form that it can be used to filter for economic relevance information generated by science.

(2) This mediating function is not to be thought of as purely passive. Technology is not a system which transforms inputs directly and mechanically into outputs. It is also a locus of creativity. Technology is the custodian of a memory – in economic terms, its capital stock of information. New information received by the technological system is integrated into this memory; and out of the new configurations that result information is made available to the economic system. But new configurations are also possible without new inputs: they may result from rearrangements of items of information already in the stock.

(3) In our particular case study there appear to have been clear limits to the extent to which information from the economic system was 'heard' as tracking data by science. To be specific: the penetrative power of the price system was very limited; the trajectory of scientific research – its path through time – does not appear to have been strongly influenced by economic considerations. Why this should have been true is a problem for the cultural historian; the answer probably lies in the particular constellation in which science and public affairs found themselves in the advanced industrial nations at the close of the nineteenth century. Whatever the explanation may turn out to be, the set of relationships shown in our case study is probably best interpreted as indicative of a specific historical conjuncture; it should not be taken as definitive or universally true. In our case study, information on market opportunities – which is to say, of economic 'needs' as ordinarily interpreted – was operative to a significant degree in the technological system, though even there it was incoherent, imperfectly understood, and evoked ambiguous and hesitant responses, except in the case of Marconi. But into the domain of science in this case signals from the price system evidently made no penetration and did not influence the trajectory of scientific inquiry.

(4) For future research in this and allied areas, particular importance might be attached to two lines of inquiry. The first is the role of what we have called 'tracking' signals fed back from one system into the other. The second is the agents and institutions responsible for the forward movement of information out of one system into another and (by implication) the selection of relevant information out of what gets defined as noise. These two lines of inquiry cover much of what seems central to the analysis of the history of technology, at least as seen by the economic historian. In both cases we are asking how information is moved, how one system 'hears' and responds to information generated in another. Most of the classic problems of cybernetics and 'general systems theory' are involved here. The economic historian has to be particularly concerned with a subset of these problems: those that relate to the allocation of scarce resources to the production, storage, and utilization of information that is new. How is this allocation problem handled? By what individuals or institutions? How effectively? At what cost? And how have these relationships varied over time and among industries and areas of scientific and technological endeavor? We have dealt in this case study with an episode in which particular individuals could be identified as 'translators' – individuals who could talk more than one language, who were accepted in more than one community, who could move information back and forth with facility. In other times and other places we might expect to find a higher degree of institutionalization and perhaps a higher degree of rationality, in the economist's sense of the word, particularly in the later stages of development of an industry when corporate organizations and their information-processing systems grow larger in scale and more bureaucratic. But the functions have to be performed nonetheless, even though identification of individuals may be more difficult and perhaps less necessary.

References

1. Joseph A. Schumpeter, *Business Cycles: A Theoretical, Historical and Statistical Analysis of the Capitalist Process*, 2 vols., New York: McGraw-Hill, 1939; Abbott Payson Usher, *A History of Mechanical Inventions*, Cambridge, Mass.: Harvard University Press, rev. ed., 1954; Jacob Schmookler, *Inventions and Economic Growth*, Cambridge, Mass.: Harvard University Press, 1966 and *Patents, Invention and Economic Change*, Cambridge, Mass.: Harvard University Press, 1972; Nathan

Rosenberg, *Technology and American Economic Growth*, New York: Harper, 1972 and 'Science, Invention and Economic Growth', *Economic Journal* **84**, 1974, pp. 90–108.

2. Erik Barnouw, *A Tower in Babel*, New York: Oxford University Press, 1966, p. 69.
3. Heinrich Hertz, *Electric Waves, Being Researches on the Propagation of Electric Action with Finite Velocity through Space*, authorized English translation by D. E. Jones, London: Macmillan, 1893; Rollo Appleyard, *Pioneers of Electrical Communication*, London: Macmillan, 1930; and Hugh G. J. Aitken, *Syntony and Spark — The Origins of Radio*, New York: Wiley-Interscience, 1976.
4. Aitken, *Syntony*, pp. 80-168; for biographical information, see W. P. Jolly, *Sir Oliver Lodge: Psychical Researcher and Scientist*, London: Constable, 1974.
5. Oliver Lodge, *Signalling without Wires*, New York: Van Nostrand, 1902, p. 45.
6. Aitken, *Syntony*, pp. 179-285; see also Degna Marconi, *My Father Marconi*, New York: McGraw-Hill, 1962; W. J. Baker *A History of the Marconi Company*, New York: St. Martin's Press, 1971; and W. P. Jolly, *Marconi*, New York: Stein and Day, 1972.
7. For comparative data on diffusion rates, see John Enos, 'Invention and Innovation in the Petroleum Refining Industry', in *The Rate and Direction of Inventive Activity*, Princeton: Princeton University Press for the National Bureau of Economic Research, 1962, pp. 307-308.
8. Talcott Parsons, *The Structure of Social Action*, Glencoe, Ill.: Free Press, 1949, and *Essays in Sociological Theory, Pure and Applied*, Glencoe, Ill.: Free Press, 1949.
9. Aitken, *Syntony*, pp. 8–20.

ON THE RELATION BETWEEN TECHNOLOGY AND SCIENCE — GOALS OF KNOWLEDGE AND DYNAMICS OF THEORIES. THE EXAMPLE OF COMBUSTION TECHNOLOGY, THERMODYNAMICS AND FLUIDMECHANICS*

GÜNTER KÜPPERS

USP–Wissenschaftsforschung,
Universität Bielefeld

1. Introduction

Combustion technology on the one hand and thermodynamics and fluid-mechanics on the other are disciplines from technology and science that possess a common core of theory. Combustion technology is concerned with installations using energy contained in fuels, for the purpose of

- processes of physical and chemical transformation of materials (for example blast-furnace, refinery, steel works),
- heating of residential or working premises,
- production of mechanical or electric energy.

In order to answer the questions connected with this, one must enlist the findings of the theory of flow, physical chemistry and heat and mass transfer. Besides the chemical data about the course of reactions and the consumption of matter and heat, we moreover need equations for the exchange of mass, substance and energy in the field of flow and mixture, as well as for the exchange of heat between the gases burned and the water-steam system. This information is supplied among other things by the

---

* This essay is based on the ms of a talk on the same subject. The part dealing with combustion technology on that occasion was prepared by Prof. Rudolf Günther of the Engler-Bunte Institute at the University of Karlsruhe. Without his collaboration, for which I here express my special thanks, this essay could not have been written. Thanks are also due to Peter Weingart and Peter Lundgreen for a critical perusal of this ms.

Krohn/Layton/Weingart (eds.), The Dynamics of Science and Technology.
Sociology of the Sciences, Volume II, 1978. 113–133. All Rights Reserved.
Copyright © 1978 by D. Reidel Publishing Company, Dordrecht, Holland.

academic disciplines of fluidmechanics and thermodynamics: the first deals with the liquid and gaseous states of aggregation of matter, especially with streams of fluids and their relation to each other and the boundary of the containing walls, the second with the fundamental approach to the theory of the properties of macroscopic matter. And, therefore, we are speaking of a common core theory of the academic and engineering sciences under consideration. Furthermore, fluidmechanics is closely related to thermodynamics. The theories of thermodynamics are not restricted to particular models of matter – they are general and model-independent. The theory of liquids and the kinetic theory of gases result if these general theories are applied to definite models. Problems of heat and mass transfer either appear as part of the field of fluidmechanics or as an independent discipline.

In the course of their history, each of these disciplines from technology and science has been repeatedly influenced by developments in the other. Thus the development of thermodynamics and fluidmechanics (hydrodynamics) in the seventeenth and eighteenth centuries was closely linked to practical problems in the domain of mining and water supply as well as to the development of the steam engine. Combustion technology was significantly influenced by developments in the natural sciences. It owes its current interdisciplinary status to advances in chemistry, thermodynamics and fluidmechanics. The chemical side of combustion technology developed first; stoichiometry, the basis for understanding chemical reactions, is over a century old, and the theory of equilibrium not much younger. The theory of chain reactions dates from the beginning of our century, while the kinetics of reaction (that is, a detailed account of the course in time of all partial reactions in a total reaction) remains largely unknown even today, because it is closely linked with the thermodynamics of irreversible processes, whose problems likewise have not yet been satisfactorily investigated. The thermodynamic branch of combustion technology is younger than the chemical one. The first law of thermodynamics is indeed more than a century old and the second law, too, goes back to the last century; but the theory of heat radiation, which is the most important form of heat exchange in combustion installations, was not formulated until the turn of the century. The hydrodynamic side is younger still. Following isolated forerunners in the 1920s, a broader development had to wait until after the Second World War. The theory of free jets, exchange of impulse and matter in these jets as well as the theory of ultrasonic beams, are based on the development of fluidmechanics and round off the subject of combustion technology. The academic disciplines mentioned appear in the standard list of subjects of today's combustion engineer, with the epithet 'engineering' (engineering thermodynamics, engineering chemistry, and so on); it is precisely the combination of these disciplines that generates essential stimuli.

Generally speaking, thermodynamics and fluidmechanics do indeed share with combustion technology the same domain of objects, namely nature, understood as including the independent course of nature as well as everything created by man (1). However, the goals are different: science seeks insight, technology builds installations. If we wish to determine the relation between technology and science, we should obviously try to do this by systematically comparing precisely such disciplines as have a common core theory, like those in our examples, but directed to different goals. In this way the rules for generating technical and academic knowledge can be found and compared, in contrast with what happens in ordinary historical analyses. In turn, these comparisons may yield a new interpretation of the historical material. A systematic comparison between science and technology thus aims at a theory for the generation of knowledge and tries to explain why knowledge systems relating to the same domain of objects develop differently and even in opposite directions, but may come together again as well (2). By choosing the disciplines of fluidmechanics, thermodynamics and combustion technology we are describing a specific set of circumstances that need not exist in all other fields of science and technology. The results obtained may therefore be of limited validity. Nevertheless, it seems to me that they are, in principle, valid for the whole domain of physics and technology. To prove this it would, however, be necessary to make the comparisons for a whole series of examples.

We base our investigation on the following theoretical assumptions: science and technology are systems of knowledge, in which knowledge is created in a continuing process of research, which is influenced firstly by knowledge already at hand and, secondly, by *criteria of relevance* that are binding on society as a whole (for example, health, prosperity, security).

By being considered in terms of the various sub-systems or functional domains of society, these criteria are given quite specific interpretations and thereby take on a specific form. The social domain to which institutionalized academic science must be assigned has, of course, no specific relevance criteria. This domain is exonerated from direct responsibility and the only criterion valid for it is 'truth': socially this criteria has been institutionalized and become the focus of an almost autonomous sphere of action (3).

Relevance criteria have a certain selective task in the process of research in that they bar some problems and questions from entering the contest. They decide what is important, what knowledge is needed. Of course, they cannot themselves organize the scientific process or direct the development of the production of knowledge, for they do not figure amongst the principles of science itself. Therefore, starting from knowledge already available, they must be transformed into determining principles of development ('Entwicklungsregulative') of the process of research. This occurs in a twofold manner: first, the demands on the knowledge required are determined (for example, description, analytic causal explanation), and second, it is decided in what fields knowledge must be made available. On this second plane the determining principles moreover determine the procedures and methods by means of which the knowledge aimed at is to be produced. The first stage of transformation sets up the *explanatory goals*, the second the *selection rules, cut-off conditions*, and *criteria of solution*. This transformation is moreover fashioned by the forms in which science and technology are institutionalized. The proportions of the variables (existing knowledge, relevance criteria, institutional factors) determining the determining principles depend on the state of development of a discipline, on national tradition and on history. Figure 1 illustrates the process of production of knowledge.

The selection rules organize the process of generating the problems, determining which questions of research are important, and which must be tackled before others. The cut-off conditions or criteria of solution indicate when the goal is reached or when a certain development may be stopped, because, for example, an unforeseen discovery makes some other development more appropriate, or, as in technological research, because further pursuit of a given question, though scientifically interesting, no

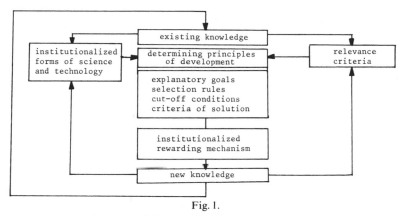

Fig. 1.

longer corresponds to the initial task. The explanatory goals at the same
time determine the procedures and methods by means of which the knowledge required is to be produced.

How best to conceptualize the relevance criteria is a matter of
methodological procedure and analytical fruitfulness. For the time being
we proceed from the notion that there is in any society a set of basic
relevance criteria which do not change in the course of time. Health,
prosperity and security, for example, are binding in a general, timeless
way. Their relative weight, and their interpretation are, however, subject
to constant change. New findings enable us to make new valuations and
therefore alter the relative weight of each relevance criterion. (Just
recently, for example, the criteria of health and security have become more
important.) Therefore the determining principles change with time, setting
in motion a process that can lead by a change of explanatory goals to a
change in the quality of the knowledge produced and so, in the course of
history, to a differentiation of the varied knowledge systems.

Determining principles can direct the research process only if they are
guaranteed as binding. This is achieved by means of institutionalized
rewarding mechanisms linked to certain forms of institution. Thus
institutions for fundamental research will reward certain kinds of activities
and institutions for technological sciences other kinds. The way various
knowledge systems differ would then be interpreted as a result of different
relevance criteria and the concomitant institutionalized reward
mechanisms.

In the following comparison of the two disciplines from science and technology mentioned above, these theoretical notions will be illustrated. The comparison is therefore made in terms of the analytic concepts that we have summarized under the heading of determining principles. Independently of and alongside historical analysis, we can thus use the systematic comparison between science and technology to indicate, by way of example, the conditions determining their relation and its continuous change. Science and technology then turn out to be knowledge systems produced by corresponding relevance criteria in analogous fashion and situated in a continuum reaching from practical empirical knowledge determined by practical goals to abstract knowledge determined by laws (4).

2. Explanatory Goals

Truth on the one hand and efficiency on the other are general relevance criteria for the domains of science and technology. These criteria do not initiate any specific developments, but they define with adequate precision what explanatory goals hold in the two domains; namely to *interpret and explain* (5) in the scientific subjects, to *interpret and realize* in the technical ones. Thermodynamics and fluidmechanics thus produce knowledge of laws, while in combustion technology the emphasis is on knowledge of rules and of construction. Of course, this often presupposes knowledge of laws. If in the appropriate academic disciplines the latter knowledge is already available, it will be applied in the engineering discipline, so that the existing equations have to be solved (exactly or approximately) under quite specific boundary conditions determined by a given installation. At the same time, existing knowledge must be transferred to the concrete technical question (6), while if the necessary laws are lacking they must be sought with the specific goal in mind, an operation that can proceed either in the academic or in the engineering disciplines. We can illustrate this in the case of combustion technology. In order to understand and therefore influence the processes occurring in a furnace installation one needs to have theoretical knowledge in the fields of chemical reactions, heat and mass transfer and theory of flow. These insights are partly available in thermodynamics and fluidmechanics, although not always at a sufficiently concrete level, so that one often has first to find the equations that will describe the special conditions in a furnace. This may not require new principles of physics but does mean that the question, as hitherto posed, must be extended to problems typically associated with the furnace. At the same time the equations must be solved under the special boundary conditions (geometry) of the furnace. If this cannot be done in a general form, one uses appropriate models or procedures (for example, calculating models of numerical analysis).

It turns out that the theoretical questions can be posed in like manner in the associated academic disciplines, but in technology they are always posed in view of their manifest bearing on application. Interest in them ceases as soon as the practical problem can be solved, even without their being clarified. The idealizing models customary in science here lose their significance and additional questions arise. Thus, for example, in the development of furnaces it is not enough to predetermine (describe theoretically) as accurately as possible the shape of the flame, the flow pattern and the course of the reaction or the radiation capacity of the flames, one needs in addition to be sure that the flame is stable (burns in the same place), that it does not oscillate, that the furnace when turned off will not be damaged by radiation from the walls of the combustion chamber or by the flames of other furnaces and that a certain domain of regularity can be reached and maintained. These additional problem areas result from relevance criteria that hold in the economic or political domain, such as efficiency, safer operation, environmental protection. The explanatory goal of the natural sciences, namely the discovery of general laws of nature but also the description of concrete objects and phenomena within it, will be strongly modified in the technological domain because of due regard to these specific relevance criteria. Analytic causal explanations are aimed at here too, but they are always related to the determining principles derived from the relevance criteria. These determining principles are an integral part of many typical questions or problem areas, such as, for example, stable flames, minimizing soot formation and laminar flow. Additional questions are defined, and general questions are made more specific.

Science and technology therefore yield knowledge of laws. Combustion technology additionally produces knowledge of rules and experience. It does not simply amount to transformation (application and connection) of knowledge already established in thermodynamics and fluidmechanics. However, these latter likewise produce knowledge of direct practical significance, although by accident or at least without intention.

3. Selection Rules

Selection rules determine the research question to be chosen from the collection of possible questions and then to be investigated. Analytically we can distinguish internal and external rules; that is, those that have their basis in the scientific or engineering system alone, and those that are anchored outside these two systems. In science the selection rules depend on the state of development of a discipline and in engineering science on its explanatory goal as well. If the formulation of general theory is not yet finished or new theories are still being formulated or derived from other theories, only those problems are generally tackled that look promising as regards a breakthrough. A widening and application of these new theories to the various domains of objects is then postponed until the real goal has been attained. Both (phenomenological) thermodynamics and fluid-mechanics are in this sense finished and completed theories. Their principles have been known for over a century. Both, therefore, develop along certain domains of objects and phenomena. Their equations and procedures are transferred to other disciplines and adapted to the physical problems that occur in them. Development is therefore arbitrary and shaped by an accidental or even externally induced interest in certain phenomena. (Thus we now have, for example, fluidmechanists in ocean-ography, geophysics and even in astrophysics. Some of their problems are: oceanic currents, the earth's magnetic field, sunspots, granulation of the sun's surface and the like. The most prominent example of an externally determined front of research is the theory of flow induced by aviation engineering, a discipline that forms part of fluidmechanics.) This interest narrows the choice of problems to be investigated, not every problem is allowed to compete. The conditions here rather resemble those in engineering disciplines: there too, it is external criteria that restrict the class of possible questions and so determine the questions to be investigated. However, even in these 'finished' theories there are blank patches, domains in which the basic equations are unsolvable for certain questions, so that new and 'better' theories must be found. Examples for this are the theory of turbulence or irreversible thermodynamics. Circumstances here are like those for the general formation of theories, there is no clear hierarchy of problems fixed purely from within science; central questions must be solved and tested before one can work out 'applications' (in technology or in physics). Too much theory is as yet missing to allow concrete problems to be solved directly. Of course, partial problems like turbulent intermixtures are interesting both as regards the setting up of a theory of turbulence and in connection with concrete practical questions (for example the spread of noxious substances in the atmosphere). That is why the formation of theory required here can proceed along the lines of these externally determined questions.

These examples should illustrate how, in our cases, the selection rules are determined internally (from the internal logic of science) or externally (from certain tasks or goals), how both possibilities exist in science and technology and how they interpenetrate. For in technology, too, the scientific core generates problems internally (for example, problems of turbulent intermixture or of the kinetics of reactions), but they play only a subsidiary role because of the strongly goal-directed nature of the enterprise. Where in science there are theoretical cores that are already finished, it is possible to develop theories in relation to objects themselves. In this the choice of problems is, however, equally restricted as in the case of externally determined applications. The difference between applied physics and technology here vanishes, for both projection onto a goal and orientation by external relevance criteria fix selection rules that no longer admit all possible questions to compete. The selection rules determine which questions are to be investigated. The questions of fluidmechanics and thermodynamics are directed to the discovery of laws that determine the behaviour of matter in the fluid states of aggregation and seek to explain the effects connected with the phenomena of heat. More particularly one seeks to know how matter under given boundary conditions will behave in equilibrium, and what processes occur on the way to equilibrium (what transport of mass and energy is associated with such equilibrating processes). This knowledge is sought quite in general or for concrete objects in nature.

Basically we meet two kinds of questions here. Firstly we are concerned with the setting up of general principles and the formulating of fundamental equations (equations of motion), namely the setting up of conservation laws, e.g., for mass, momentum and energy, and the equations derived from them for certain effects. Secondly, we have more special questions concerned with the concrete solutions of these equations above all in the 'applied domains' of these disciplines (for example, oceanography, geophysics, astrophysics). Such 'applications' of the general principles are by no means trivial tasks. Often here too, we must begin by deriving the 'correct' equations (namely those that describe the physical state of affairs of interest) and then find approximating procedures for solving them.

An example of the first kind of question is statistical mechanics, whose goal is to derive all equilibrium properties of a macroscopic molecular system from the laws of molecular dynamics. Thermodynamics is a partial field of this domain in the sense that special assumptions about the properties of atoms and molecules and their mutual interaction lead to the kinetic theory of gases which can be construed as a microscopic foundation of traditional phenomenological thermodynamics. Between the two groups of questions lies the field of turbulent flow. This phenomenon is, indeed, described in the basic equations of fluidmechanics but because they are unsolvable we can make no concrete statements about it. Here we must find a new description within whose scope concrete questions can be answered. For example, what is the heat transport in a turbulent stream given the temperature gradient and the geometry of the system?

To the second kind of question belong the previously mentioned cases of 'applications' of the general theoretical cores of thermodynamics and fluidmechanics to concrete objects in nature. Here the goal is to explain phenomena to be observed in quite different areas of nature. As an example we may take the whole of fluidmechanics where we examine the behaviour of flow of liquids and gases under quite special conditions as to geometry (e.g., of the containing walls) or order of magnitude of density, temperature, velocity and the like. The main questions may concern the law-like behaviour of these flow patterns themselves or the equilibrating and diffusion processes provoked by them. Here we are still concerned with basic problems, although the systems are, of course, progressively more specialized. Thus one now examines, for example, vibrations by means of ultrasonic flow, the spread of sound in flowing media, flows in real dilute media and molecular interactions in attenuated jets. In this connection one even investigates domains between parts of classical physics (fluidmechanics and the physics of atoms and molecules). To this extent we have, alongside concrete 'applications' of thermodynamics and fluidmechanics, an effort towards setting up a theoretical core of modern fluidmechanics valid for these domains.

Combustion technology is concerned with the use of energy contained in fuels for externally determined ends. This, too, raises two kinds of questions. The first aims at the construction and empirical investigation of installations and apparatus, the second at the physical and chemical processes taking place in these installations. The two types of questions do not occur in isolation but are closely related to each other. Empirical knowledge of construction builds on the available knowledge of scientific laws, and that in turn is sought for the sake of a better understanding and influencing of the processes occurring in the installations to be built. In the foreground of interest there is always the joint action of many partial processes in an installation. In combustion technology it is the joint effect of mixing and firing the fuel, and the course of combustion with the attendant problem of the release of heat. The table uses the firing of a boiler to indicate the questions linked with certain functions and their corresponding items of plant.

Partial Processes in Firing a Boiler

Heating of the fuel oil to vaporization temperature	Oil pre-heating unit
Producing the vaporization pressure	Pump
Pre-heating of the air admixture	Air pre-heating unit
Transport of the air admixture	Air blast
Vaporization of the fuel oil	
Evaporization of the oil drops	
Mixing of oil vapour and air stream	Burner and combustion chamber

Ignition of the mixture	Burner and
Burning of the oil in many reaction stages	combustion chamber
Heat exchange between combustion gases and water or steam	
Disposal of exhaust gases	Suction pump and chimney stack
Dilution of exhaust gases with surrounding air	(Smoke plume)

To solve such problems combustion technology needs to know the laws of thermodynamics and fluidmechanics. In our example we therefore have a common 'core of theory'. How close the links with thermodynamics and fluidmechanics are is shown in the problem of turbulent flames. The basis of a theory for the course of reactions in turbulent flames is formed by the most varied hydrodynamic theories of turbulence. The oldest of these is due to Prandtl, who tried to set up an analogy between the behaviour of vortices in a turbulent field and the molecules of a gas. Later Taylor applied the methods of mathematical statistics to turbulent processes. However, there is not, to date, a satisfactory solution for this domain of phenomena in flowing media. Hydrodynamic theories of turbulence are applied in the engineering theory of flow to gas jets that are to spread in stationary surroundings. Combustion technology then applies available information of free jets to jets with reaction (combustion), studying how in a field of turbulent flow the accumulation of mass and the mechanism of local exchange is modified and how the reaction fields range themselves geometrically into the vortex system.

At this point combustion technology is closely linked with the frontiers of research in thermodynamics and fluidmechanics, without losing its relation to practice. Thus one enquires not only quite generally into the spatial distribution of turbulent quantities of exchange for impulse and mass in the jet flame of a certain fuel, but one looks for concrete possibilities for influencing the course of burning, the geometry of the flames or the reaction of a fire box in the course of burning. Here there will be few theories and often it is empirical knowledge that enables such problems to be solved. The connection becomes even more obvious if one considers the development of fluidmechanics which was determined by the constant accretion of new effects (such as, for example, sound and burning in free flame jets). This led to the examination of questions closely connected with combustion technology. Of course, it was not concrete external requirements that have led to the adoption of such research questions, but a discipline with a fairly completed core of theory 'automatically' received a wider scope of application. To that extent, this scope of application emerges accidentally and is not determined from within science. In contrast with the more theoretically oriented questions of thermodynamics and fluidmechanics so far described, there are additional questions in combustion technology concerning the theory of construction, the technology of measurement and regulation energy, and problems of working materials. In growing measure, environmental questions come to the fore as well. These problem areas closely linked with the realization of a project relate above all to empirical knowledge which must replace or supplement the missing knowledge of laws. This situation will not essentially change so long as we do not succeed in developing theories for complex systems. (To the know-how in the building of installations there corresponds, in thermodynamics and fluidmechanics, the knowledge of calculating techniques (solvability and convergence of certain procedures).)

The way in which selection rules operate has been discussed in relation to the various typical questions. Since these rules follow from differing relevance criteria, a clear difference between science and technology should become visible here. However, the examples showed great similarity. In combustion technology the mainly externally determined selection rules furnished 'theoretical' questions as well, while in thermodynamics and fluidmechanics an autonomous generation of problems led to questions with a clear practical bearing. The differences between 'internally' and 'externally' derived selection rules vanish in the questions they generate. The reason for that is to be seen in the particular structure and state of development of the knowledge concerned.

4. Treatment of Problems (Methods and Modes of Working)

So far we have tried to compare disciplines from engineering and academic science in terms of their explanatory goals and the way they generate problems. We must now proceed to carry out a systematic comparison in terms of the methods used in treating problems and of the ways of working typical for the various domains.

(a) *Experimental procedures*

In both domains, as a rule, processes either cannot be fully described theoretically or the systems of equations describing the processes cannot be fully solved. This holds for processes in nature as much as for those occurring in machines or installations. Thus, relatively simple processes in nature, like, for instance, the flow pattern round a given body in a liquid, are generally not describable. The behaviour of turbulent flow cannot even be formulated exactly; processes of energy exchange in 'realistic' systems with heat being convected, radiated and dissipated cannot be calculated. In the first case the exact equations are unsolvable, in the second, there is no theory and in the third, the system of equations becomes too complicated when all effects are taken into account and therefore again unsovable.

The same holds for complex processes in machines and installations. There may indeed be theories, equations and solutions for this or that partial process, but their joint action escapes description and therefore solution. In such domains one, therefore, must conduct experimental investigations, in order to obtain the lacking data and to test theoretical approximating methods and models. Experiments are always closely linked with special measuring techniques and procedures, because isolated quantities are measured as to dependence on certain parameters. Every field must rely on existing procedures which it adapts to special conditions, failing which it must develop its own procedures. In this way one develops procedures for measuring temperature and density partly by optical means and partly with probes. For this, one often requires a precise theoretical grasp of the 'elementary processes', for example, in order to be able to infer the temperature or density of light from its scattering or the polarization.

Each discipline develops its own range of instrumental procedures. In combustion technology this may be methods for measuring radiation, turbulence in flames and the speed of flames. Methods developed in physics for wind gauge measurements using lasers and Doppler effects. Raman and mass-spectroscopy, interference and scattering spectroscopy, are adapted to the conditions of combustion technology. Measuring techniques developed in this way are, of course, highly important in thermodynamics and fluidmechanics as well, so that as regards the development and use of measuring procedures there is no difference between the two technical and scientific domains. In both, we measure local or mean temperature, concentration, velocity, heat flow, radiation, oscillations, noise and the like.

Moreover, experiment has the same function in both domains. In combustion technology it is the installation itself that is the experiment. This is given so that here we can speak only of an empirical way of obtaining data: the application and development of special measuring procedures are, therefore, in the foreground. By contrast, in thermodynamics and fluidmechanics, experiments are 'excogitated' and correspond to a representation of natural processes with many parameters on an arrangement with few but highly testable and adjustable variables. In both cases experiments serve to clarify certain phenomena and the connection between different quantities: the task is to make the influence of some factors visible while suppressing that of others, which thus amounts to asking nature or an installation an idealized question. For even in a technical installation, the joint action of many factors makes it impossible to gain a precise picture of all function.

Often too, one must examine effects that occur in large and complicated installations or in inaccessible objects in nature, or in rather large extensions of space and time, so that, failing a commensurate effort, they escape direct measurement. Here we must rely on models that simulate nature or a technical installation. The same holds for processes that are so complicated as to require a preliminary reduction to partial processes in a model, before they can be studied further. In combustion technology, such geometrically similar but highly scale-reduced models of a machine are called 'physical models'. The goal of making models is to save money, thus to avoid the financial risk involved in experiments or to create conditions for measurement at all. The basis for this practice is a theory of 'similarity' developed in fluidmechanics, which states whether and under what presuppositions such a model behaves like nature or a large installation. In this process, certain characteristic numbers indicate how to transfer the measurements from the model to the actual question. In combustion technology one speaks of 'cold' models, if in a combustion installation the chemical reactions during burning are either left out of account or are simulated by mixing two gases or liquids.

Of course, the models become progressively worse, that is their relation to reality becomes more and more slender the more complex the processes to be simulated. Often too, it is the special nature of the question that prevents its study in the model, so that one must conduct investigations on the object itself. This holds whenever events are governed by non-linear effects. For example, resistance to flow by given bodies can be studied only in a wind tunnel. In this way we can sometimes use models to treat questions of principle, such as the influence on a flow pattern of changing the geometry of the boundaries (of a flow channel or a combustion chamber), while the properties of new furnaces can be examined only in actual size, even if we have to reckon with cross-sections of 26 m² and oil throughputs of 5 t an hour.

(b) *Theoretical models*

So far we have spoken only of experimental investigations and the role of models in them. However, there are also theoretical questions in both domains. Their goal is to reveal and formulate general connections, but also to give quite detailed theoretical forecasts about the behaviour of certain quantities for given boundary conditions. In both cases models (here 'mathematical' ones) are indispensable. Often the fundamental equations are unsolvable for quite simple processes (for example, in fluid-mechanics the general problem of three-dimensional time-dependent flow); or, if all important effects and partial processes are taken into account, we obtain complex systems of equations which are either unsolvable or require a disproportionate effort for solution. An example from combustion technology is the exchange of radiation between a flame and the walls of a combustion chamber. Thus one is forced to idealize problems and neglect 'side effects' both in setting up basic equations according to the conservation laws of mass, momentum and energy, and in their solution. Of course, we often know very little about the joint action of all neglected side-efforts, so that such approximations remain rather 'academic', having little resemblance with reality. However, this is less important in academic disciplines than in combustion technology (where the aim is to solve concrete problems). In thermodynamics and fluid-mechanics a result holds even if it was obtained for frictionless fluids, that is without dissipation (due to frictional heat), so that it does not necessarily answer a concrete question concerning a definite liquid.

In particular, both in thermodynamics and fluidmechanics and in combustion technology one neglects effects like friction, temperature distribution, and conductivity, while assuming a certain geometry (plane, linear, axially symmetrical) and making special presuppositions about the nature of interactions (for example, purely mechanical, without chemical change of substance, like rigid spheres, without heat exchange). However, whereas in combustion technology we must reach an approximation to reality (the correctness of findings from the model are tested at the latest when an installation is built and put in operation), this is not the case in thermodynamics and fluidmechanics. Here we are not compelled to solve, for example, problems about real liquids under realistic boundary conditions.

In the formation of models, the criteria for solution to be discussed below play an essential role. They ultimately define the admissible 'loss of reality' and thereby determine which models remain admissible. In practice, it often happens that mathematical and physical models are used in combination. This holds for the academic disciplines of thermo-dynamics and fluidmechanics as well as for combustion technology. A peculiarity of the latter is that, in order to overcome uncertainties of scale between laboratory experiment and large installation, one inserts a half-empirical experimental installation or pilot plant. This too, is an attempt to keep risks and costs within defensible proportions.

5. Cut-off Conditions and Criteria of Solution

Just because natural sciences must reach a late stage before they become sufficiently concrete to be applicable to concrete technological questions, technology itself must often produce the missing knowledge. This it does by means of the same methods and modes of working as are used in the academic disciplines, although it is, of course, not quite as unfettered. More 'selective' determining principles prevent developments too divergent from external requirements. The role of the selection rules has been discussed above. They are supplemented by cut-off conditions and criteria of solution which, in academic sciences, stipulate when we may leave the indicated line of development before the original goal has been reached and turn in a new direction. This is important above all in the discovery of unexpected effects that open up a new and unforseen frontline of research before the old one has led to the original goal. In engineering research, criteria of solution have, so to speak, the reverse function. Here they determine when the solution is reached and research stopped even if theoretically (in the scientific context) it is likely to lead to further interesting results, or the phenomena under study may not even be extensively clarified and understood. In academic science, a problem counts as solved if the underlying principle has been formulated (for example, first and second law of thermodynamics) or when for a certain domain of objects and phenomena the valid 'equations of motion' have been found and the important general properties of their solutions are known (for example, existence, stability, behaviour in space and time). One proceeds from certain boundary conditions; that is, from notions about the relevant geometry, about values of parameters at boundaries or 'at infinity' and from assumptions about spatial and temporal behaviour of physical quantities (for example, constant in space or time or both, monotonic increasing, negligibly small).

Starting from quite simple systems and models one gradually tries to come closer and closer to the true state of affairs in nature. In this, of course, one does not examine the lawlike structure of special systems: rather, the question is one of principles valid for all systems of the same kind. In this process a fairly idealized system takes on the role of a test system with which all the others are compared. The test system constitutes the zero line, as it were, it represents the 'undisturbed' ideal system. This shows that in academic sciences the range of solution is fairly widely stretched. Every proposition that, starting from certain presuppositions, makes a provable assertion is a solution. The attendant idealizations often go very far indeed, physics become 'mathematized'. An example will illustrate the point.

Let us consider the problem of the flow of liquids or gases. If in the equations of motion we neglect friction (viscosity) and thermal conductivity, we obtain a theory of 'ideal fluids' in which we can formulate Kelvin's vortex theorem that the vortex strength remains constant along a streamline. We can discuss special forms of disturbance marked by an additional condition satisfied by the velocity vector, leading to a rather simple equation of motion. The theory of potential flow is an example. Here again we can formulate general laws about the properties of solutions to the potential equation. In all this it is perfectly clear that there are no frictionless media in nature, so that such theories fully describe conditions only where friction is not the dominant effect. If, however, the small but existing frictional effect is nevertheless to be taken into account, one must find new equations that are much more complicated. This shows that in this case one speaks of a solution even when only a very simplified ideal case has been treated and solved.

In engineering science the case stands otherwise. Here we can speak of a solution only when the installation is built and functions. This does not require that everything is understood; nor need the plant run under optimal conditions. Where we are concerned with understanding how certain things function, the problem is solved (as in the example from thermodynamics and fluidmechanics) when the connections looked for have been clarified for a typical (model) case. Nevertheless, model solutions are important only if they facilitate or promote the realization of an installation. Variations in parameters remain unconsidered so long as there is no prospect that they might lead to new theories that could be the basis for improving existing installations or building new ones. This stops further autonomous development of the theory.

6. Conclusion

In this essay we have attempted a systematic comparison between an engineering discipline and the corresponding academic disciplines, guided by a set of concepts which together are conceived of as determining principles directing the production of knowledge.

As regards the domain of objects of science and technology, we could indicate no difference, since in principle one cannot distinguish between nature and 'artificial nature'. In both domains the same nature is involved and the same natural laws are valid. Processes in nature and in an installation have about the same degree of complexity. They are composed of a great many individual processes and correspond to a joint action of the most varied effects. Nor does the difference in explanatory goals (to interpret and explain in science, to interpret and realize in technology) affect the methods and procedures by which the respective knowledge is produced. The development and use for instruments for observing and measuring show just as little difference.

The difference in explanatory goals leads, however, to differences in a partial domain of the determining principles, namely the selection rules. Whereas in engineering sciences, the external determination of goals limits *a priori* the number of questions admitted to compete (only those are admitted which need to be clarified for the attaining of the goal), in the domain of natural science, at least, where one is concerned with the general formation of theories, the number of problems is not limited from the outset. Of course, this does not apply where in natural science one 'applies' existing cores of theories to domains of phenomena and objects of nature. Here conditions are similar to those prevailing in engineering sciences. Here too, the goal allows a certain phenomenon to be explained and questions to be classified into 'important' and 'unimportant', thereby reducing the class of admissible questions.

The difference in explanatory goals further leads to different concepts of solution; in the natural sciences a solution for an idealized model case is enough. This model case symbolizes 'pure' nature, reality is a disturbance of this state and amounts to a 'dirt effect'. In engineering sciences a solution is reached only when the desired installation has been realized. A reduction to problems of a 'pure' machine is here not allowed or remains unsuccessful. In other words, the domain in which the knowledge produced is valid differs in size. To the knowledge of laws in natural science with its domain of general validity, there correspond the knowledge of rules and construction in technology where the domain of validity is confined to a single installation.

What remains open here is the question *how* the determining principles direct developments in the various disciplines. One explanation of this would be that keeping to the specific selection rules and cut-off conditions is secured by means of certain institutionalized forms with different rewarding mechanisms in each case. These mechanisms then give special weight to originality and ability to abstract, or to pragmatism and a 'sense of reality', according to the institutional context.

To analyze these connections one would have to examine the various institutionalized forms of knowledge production and observe their effects on the process of research. This demands extensive empirical investigation. At any rate, we know from the past that when science became different from technology this development was linked with a difference in the institutionalized forms of the production of the two systems of knowledge (7).

Since institutional factors are often difficult to sever from cognitive ones, a difference between science and technology is simulated on the cognitive plane as systems of knowledge whose differences can be explained by the different ways in which specific determining principles are institutionalized. In this sense science and technology cannot be different in principle. They lie on a continuum reaching from practical empirical knowledge to abstract universally valid knowledge of laws. Their mutual relation constantly changes, being determined both by the specific relations of the corresponding units of systematic (here engineering and theoretical physical) knowledge and the way in which they develop at a given time, and by the continuous changes in society.

Notes

1. F. Rapp, 'Technik und Naturwissenschaften – eine methodologische Untersuchung', in H. Lenk and S. Moser (eds.), *Techne, Technik, Technologie*, Pullach: UTB. 1973, p. 289.
2. P. Weingart, 'Wissenschaftssoziologie und Technik. Die soziologische Analyse wissenschaftlicher und technischer Handlungsorientierungen', in P. Lundgreen (ed.), *Zum Verhältnis von Wissenschaft und Technik. Erkenntnisziele und Erzeugungsregeln akademischen und technischen Wissens*, Science Studies, Report 7, USP Wissenschaftsforschung, Bielefeld, 1976, p.22.
3. N. Luhmann, 'Selbststeuerung der Wissenschaft', in N. Luhmann, *Soziologische Aufklärung*, Köln, Opladen: Westdeutscher Verlag, 1970, p. 246.
4. Weingart, *op. cit.*, p. 27, Note 2.
5. Interpretation here corresponds to the setting up of phenomenological theories which do indeed enable one to make analytic causal statements, but without explaining what happens. That is achieved by 'microscopic' theories, which take into account all elementary processes.
6. To apply means: the equations for a given question are already available; to transfer means: from general principles we must first set up the equations that hold.
7. cf. the contributions of G. Böhme, *et al.* and P. Weingart in this volume.

PART III

SCIENCE AND TECHNOLOGY
IN THEIR SOCIAL CONTEXT

TECHNOLOGY ACADEMISED
Education and Training of the Engineer
in the Nineteenth Century

KARL-HEINZ MANEGOLD

Technische Universität, Hanover

In the late 1830s, the aims of economic and industrial development in Germany became more generally evident. With this in mind, Robert von Mohl, in a kind of sociological analysis, observed that within the framework of social groups signs of a new profession were becoming noticeable that could not be adequately described in terms of traditional categories of social classification, such as the distinction between those who were educated and those who were not. This, he held, was the growing group of machine builders now so essential, directors of large technical enterprises and factory managers, who could no longer be reckoned amongst artisans who serve society only with their hands; though at the same time this new group could not be classified together with theoretical scientists who were academically educated (1). By that time, Mohl's observations had long ceased to be isolated. Some years previously Friedrich List had stipulated in the political lexicon of Rotteck-Welcker, the bible of early liberalism, that the 'higher professional technicians of Germany' should unite and organise themselves in the way that German scientists and medical men had already done (2).

Until the mid-nineteenth century the term 'engineer' was by no means in general use to designate members of this somewhat vaguely defined professional group of which the only initial fact concerning their social history was that they were above the artisans. Previously it was the military meaning of the term that was predominant both in literature and in ordinary usage. Following Dutch and, above all, French practice since the early seventeenth century, it had likewise established itself in Germany in connection with the development of the army. Since then, it was mainly

Krohn/Layton/Weingart (eds.), The Dynamics of Science and Technology.
Sociology of the Sciences, Volume II, 1978. 137–158. All Rights Reserved.
Copyright © 1978 by D. Reidel Publishing Company, Dordrecht, Holland.

used to refer to the military engineer and the builder of fortifications. Just as the concepts of 'technology', in the sense of national industrial technology, and 'technologist', did not become widely used until the twentieth century (3), so likewise civilians, who had mastered the arts of construction and machinery, were rarely called 'engineers' (4). In France the organisation of units of military engineers since the beginning of the eighteenth century had produced a more precise definition of the profession; at the same time this led to a first separation and specialisation in engineering, and, according to the political, military and mercantilist needs of the Ancien Régime, to the foundation of the famous specialist schools of France (5). In many respects the École Polytechnique in Paris, founded in 1794, still belongs to the tradition in which these institutes had been set up. Since, however, it was a result of the revolution, it marked a turning point for the training of engineers and the development of technology based on mathematical scientific principles; a change that pointed forward to the nineteenth century, especially as seen from Germany. There could be no doubt about the academic status and the scientific estimation of the institutes of the French state, nor about the very high social prestige of those who graduated from them and without exception entered the army or the civil service. Since the mid-eighteenth century the French rendered the concept of 'engineer' more precise by the attribute 'civil', a term used also to refer to a non-military profession that need not depend on certain training qualifications. At the same time in the industrially progressive setting of England there developed a non-military professional group of men who held no office; from 1770 they organised themselves into professional associations such as the Society of Civil Engineers and a little later the Institution of Civil Engineers. Here, early efforts were made to delimit and define the new profession, giving a specific picture of it, although at this stage it could, of course, not yet be described as academised. This was no longer a question of military technology, but rather of a comprehensive domain of technical industrial problems entirely dedicated to progress.

In Germany no such organisations existed until about 1850, and only when they arose after that date did the term 'engineer' gradually establish itself for those civil or 'higher' technicians who possessed knowledge going beyond the purely empirical skills of the artisan, or who had gone through a corresponding course of training and occupied a leading position in the hierarchy of an industrial concern or indeed owned a factory as engineer entrepreneurs. The profession of civil engineer as an independent occupation, such as a consulting engineer, does not appear as a separate phenomenon in Germany until the 1870s, in contrast with the older profession of architecture. On the model of the English associations of engineers, a number of such bodies were formed about 1850, mostly locally; the first one in Vienna, not by accident. Moreover the periodical 'Der Civil Ingenieur', published from 1848 and since 1853 re named 'Der Ingenieur – Zeitschrift für das Ingenieurwesen', one of the earliest technical scientific periodicals in Germany, showed that engineers were beginning to grasp more clearly what united them in their professional domain and in their view of themselves in society; or, using Mohl's formulation mentioned above, to become more aware of their growing into a 'profession'. In the first issues of this periodical the question 'what is an engineer?' was repeatedly raised, while Max Maria von Weber, one of the leading railway experts of the 1850s, tried to give an entirely valid and characteristic 'definition of status'. He explained that this new profession stood "in the centre between the professions of scientist, artist, business man and artisan, related to each of them but distinct from all in virtue of its own peculiar element, regarded by other professions as an inconvenient newcomer" and occupied "a world position" whose "awkwardness and insecurity" was quite out of proportion with the weight that it represented in the current and future "life of nations" (6). Above all, however, Weber here firmly established that the position of engineers could be properly defined only by education and training and corresponding examinations: only thus could the profession be given a secure foundation and future. So he realised that training qualifications and educational criteria must be constitutive elements both for professional and for social groupings. How confused and insecure a picture the profession actually presented emerges from the fact that the early members of the Verein Deutscher Ingenieure (Association of German Engineers, founded 1856) were called by such diverse professional names as engineer, technician, teacher, professor, technical director, but also, more modestly, master mason and master carpenter. The VDI was a fast growing professional body which soon had the largest membership of engineers, and aims quite comparable with those of the Society of German Natural Scientists and Physicians, especially on the plane of national politics: their object was, in programmatic style, 'the close collaboration of intellectual forces in German technology for mutual inspiration and further development in the interest of German industry as a whole" along with the aim of allowing "technology as a co-equal intellectual attainment with science and art" to gain the same influence as the attendant ambitions for social advancement of its representatives (7). Questions about the education and training of engineers henceforth remained a constant and basic part of the Association's interests.

Just as comparable organisations in Austria had arisen from the ranks of former students of the Vienna Polytechnic Institute, so, too, did the VDI from the association of former graduates of the Berlin Institute of Trades. Indeed, the polytechnic schools play the decisive role in the genesis of engineers as a professional group. Since mid-century, in Germany these schools had won a solid position as institutionalised education centres for engineers, while in comparison the older mining academies, the earliest and most important institutions of technical scientific teaching, had been of only limited significance. We cannot here enlarge on the rise and early history of these institutions, which in direct connection with the industrial revolution had quickly developed into the most important branch of scientific organisations founded by the state (8). Some remarks on this subject are, however, necessary in connection with our more specific subject, precisely in order to bring out the internal and external problems and tensions that reveal the process of their becoming academised.

Just as all scientific and teaching institutions stand at the nodal points of manifold interconnections, so likewise institutions of technical education were determined from the outset as much by interdependent elements of economics, social, intellectual and educational history and of politics as by technical and scientific developments themselves. In early nineteenth century Germany, the motive for founding and the aims set for polytechnic schools or their precursors, were inspired by the wish to industrialise and were, therefore, marked by the problems of starting in a country, which industrially was a follower. Of all the forces important for determining these foundations, science and education were regarded as the most vital instrument for a new economic policy. "Technical scientific progress almost entirely occupies the role of pacemaker for economic development", as Karl Friedrich Nebenius had declared in 1833 on the occasion of the decisive reorganisation of the polytechnic school at Karlsruhe (9). In Prussia, Beuth, the founder of the Berlin Institute of Trades, stated that "where science is not introduced into the trades and made into the basis of production, there can be no progress" (10), while his colleague Kunth condensed the awareness of such problems into this much quoted formula: "Against the danger that the efforts of the more advanced West European industrial countries will always limit us, what help the state can give is summed up in one single word: education" (11). Mobilising science and education meant above all technical education, which was the essence of the aims repeatedly advanced here. The recognition that science and education can be effective economic factors (the pedagogic and scientific policies of the seventeenth and eighteenth centuries had already established the long tradition of this way of thinking) emerged more widely and consistently at the threshold of the industrial era in Germany and became a corner-stone of a new state policy of promoting the trades. Within this framework the goals of the polytechnic schools founded in Prague in 1805, in Vienna in 1815 and soon afterwards in Karlsruhe, were essentially parallel, just as in a certain, perhaps more ideal sense, they based themselves on the École Polytechnique in Paris. Above all, however, it was the Vienna Institute that until mid-century remained the unattained example for all later foundations. Even for Germany proper, Vienna became the seedbed of a whole generation of the new 'polytechnicians', the first institute of technology in the spirit of the nineteenth century as regards organisation, scope and level (12).

Following the insight of the École Polytechnique in Paris, that all technical work had a common basis in method, mathematics and science, the idea of an institute of technology as clearly formulated in Vienna was consciously separated and delimited from the university. According to the definition of the technologist Prechtl, who founded the Institute, the nature and unity of technology here formed the inner principle of organisation, with the explicit demand for status equal to that of the university. Conceived as centre of all efforts in the field of technology and the trades, the form of an institute of technology here found its own justification in the claim of being the new type of college for the technical services of the state and, in particular, for what was then described as the "class, important to the state, of higher industrialists, entrepreneurs and businessmen", namely the industrial and commercial bourgeoisie; not in the sense of a specialised trade school but as a university of the technical sciences, a college for an independent sphere of life. It is worth remarking that in the pioneer foundations in Prague and Vienna it had been the nobility there that became active; for, in contrast with their colleagues in Prussia and the rest of Germany, they had taken considerable interest in industrialisation. This too, no doubt, was responsible for the fact that in the Austria of that period technological and scientific studies enjoyed much greater social esteem. In a way, the dynastic conservative elite, compelled by the scramble for political and economic power, corresponded to a model of partial modernisation.

At the core of the organisational plan there was what Prechtl, its author, described as "the peculiarly rational technical method", the attempt to arrive at a specifically technical and scientific domain that was as sharply contrasted with 'pure' as with 'applied' science and 'mere empiricism'. The task then was to reach an autonomous area of scientific technology in which it should become possible to reconcile scientific theory and the empirical practice of the trades; that is, in the conviction that technical science was not the same as applied science, in opposition to the views of the École Polytechnique in Paris. Thus a kind of academising programme and an emancipatory goal were clearly formulated together, and their realisation became a central aim of nineteenth century institutes of technology, although the goal was not reached until the end of the century.

Here it must be mentioned that at the same time as the polytechnics were being founded, the idealistic humanist conception of the university developed a new concept of science and education. After the foundation of the university of Berlin, this development soon gained ground. The opposite view, against which this new university established itself, was precisely the more technically oriented enlightenment conception of the institute as a place of science aiming at application and utility in civil, economic and industrial life, and above all of the specialist trade school designed to give professional training; that is, against the traditional position of the polytechnic schools. This is not the place for enlarging on the humanist conception of education and science, according to which the university was defined on the ideal basis of 'pure' science free from any outside purpose and pursuing only its own questions, unwilling to enter into direct contact with practical affairs. Even if this conception was never preserved or realised anywhere, it still seemed quite paradoxical that at a time when people declared science to be an essential basis for a practical sphere that needed changing, this ideal guiding notion of science became the determining principle for institutionally separating science from technology; the more so since the natural sciences, too, soon established themselves successfully and in the course of the century rose to an internationally leading position, while maintaining a view of themselves as a kind of scientifically oriented new humanism of 'pure' science, a tendency that soon became more and more emphasized rather than the reverse. The course was thus set from this side too, so that the question as to a higher technical education was answered throughout Germany in favour of independent technical institutions. After some vacillation and several unsuccessful attempts at integrating the technical subjects into the university were made science and technology were unified in the one institution.

At the same time this raises the problems of the debate that lasted throughout the century concerning the dual path which constantly found new fuel in the increasing discussion of university reform from the 1820s onwards, especially in connection with the criticism of the university as being too remote from practical affairs (13). The technical institutions, in their aims at advancing and academising themselves, were drawn into the perennial controversy between realism and humanism; that is, into direct contact with the conceptual and latterly organisational antagonism between 'pure' and 'applied' science, theory and practice, education and training, intellect and industry, culture and civilisation or whatever other name was given to these antinomies in the struggle for the emancipation of technology, and they were stressed not only by an anti-technical critique of culture. The place of technology had to remain outside the university which was hostile to economic affairs or at least to technology and industry, but its development had nevertheless to proceed in constant assimilation to and adoption of academic forms which alone were socially important and set the norms. The institutional gap between science and technology resulted from many external causes but nevertheless seemed to correspond to immanent regularities, for it proved lasting and became the prototype of the organisation of science in almost all the subsequent industrial states, but also in the old industrial centres.

Even before industrialisation actually gathered momentum, the short decade from 1827 to 1836 following Prague, Vienna and Karlsruhe, saw most states in the German Confederation founding higher institutions for technical education for the trades and the technical services of the state. At first these schools were of variable level. In almost all cases the initiative for the foundations had come from a higher civil service that was progressive and liberal as regards state education. This often led to the typical confrontation between a conservative administration of liberal progressive views in economics and a politically liberal middle class with conservative economic ideas. By mid-century all these institutions received the status and designation of polytechnic schools, while at first remaining rather diverse as regards political and social esteem. In Prague and Vienna the politically conservative monarchist forces put their trust in scientific and technological progress and emphasised that the foundation of polytechnic institutes and therefore the furthering of non-political scientific studies was at the same time able to keep a young elite away from dangerous political ideas and activities, a twofold argument in the spirit of preserving the existing ruling class. By contrast, in Germany conservative opinion pointed to the fact that the École Polytechnique was an acquisition of the French revolution and argued that 'polytechnology' was close to the radical movement, trying to describe it as 'revolutionary science'. There were indeed warnings against the danger of 'polytechnicism' and 'Americanism' even in the early liberal camp, but mostly the liberals regarded rational technology as a liberating power and praised the polytechnic schools as 'citizens' universities', universities of 'real studies' or of 'technology', with the attendant demand for comprehensive development and full equality with the university as regards status (14).

In the reform movement of 1848 such demands had reached a peak: corresponding to civic equality there should be an equality of status for those institutions and disciplines that notably contributed to the rise of the economically active middle class. The initial objects of founding the polytechnic schools had often mentioned these as the summit within an articulated system of lower and middle institutions for technical education, or non-classical ('real') secondary schools, alongside the progression from schools for general education to university; nevertheless, the polytechnics had all been founded as finishing stages ahead of the appropriate infrastructure. The founders had started from the fact that what was needed first and foremost was highly qualified technically educated men, the formation of technical cadres, the 'officers of the army of the trades' as it was often called; given the gap between the level of polytechnic training and the actual general state of production in the trades, for a long time there were few if any jobs in industry for these men. Indeed it was they who were meant to help to create the appropriate conditions of demand, in conformity with the goals of the state's efforts to further the trades. Thus at first it was the technical services of the state that engaged staff qualified by state examinations requiring precisely defined knowledge and training. The consequent link with the norms and valuations in the careers of civil servants and their entitlements was thus a vital determining factor in the academic development of the institutions. In a profound essay (15), Peter Lundgreen has recently pointed out that as regards the connections between technical education and economic growth that have to be studied here, it is difficult actually to measure to what extent the amount of technical education produced meets the demand of the industrial trades. On the institutional side we can indeed determine only the history of supply, namely the output of technical education as such. There is a total lack of specific enquiries here, failing which we still know very little about the demand, namely how this output is used in industrial technology. This holds throughout except in chemistry and chemical industry which are a special case. It was not until the 1860s and 1870s that rapidly growing numbers of graduates from institutes of technology found possibilities of employment in private industry that corresponded to their qualifications.

The further development and expansion of the institutes and their teaching was indeed tied to the practical needs of industry and active technology but in equal and even growing measure it was influenced by the goal of scientific growth and self-contained progress of technology as such. This goal was inseparably embedded in the ambitions of the institutes for scientific and social advancement, with growing emphasis on equality and security of status and on social integration. In this complex network of factors the process of becoming academised certainly turned out to be the determining motive, with lasting influence on the relation of science to technology within the teaching programme and the consequent tensions between technical subjects and the natural sciences, as well as between technical sciences and active technology. Redtenbacher (16), director of the polytechnic school of Karlsruhe and one of the founders of scientific mechanical engineering, and Karmarsch (17) of Hanover, the leading German technologist at mid-century, had stressed that one had to reach an equilibrium between mathematical scientific knowledge and methods on one side and the knowledge and ability in constructive technology on the other, which latter they saw as a freely creative activity, emphasizing that mathematics and mechanical principles alone cannot lead to a construction actually carried out. However, since the 1860s an increasingly theoretical approach to technical problems came to the fore. The scientification of technology meant, in particular, a preference for an often one-sided mathematical treatment while technical practice and empirical insight were carefully pushed back as a guarantee for academic respectability. It had to be proved that technical inventions were scientifically teachable and therefore proper topics for institutes: thus stated Franz Reuleaux, at one time amongst the most influential engineering professors in Germany. According to him, one had to apply academic criteria to prove that in the technical sciences the same intellectual operations could be carried out as those used in research conducted in universities, so that technology was equal in rank even when measured in terms of the standards of scholarship; indeed he emphasized the difference in forms and goals between institute and practical life (18). The comprehensive theories of machines that were linked with names like Grashof (Redtenbacher's successor) or Reuleaux are mentioned only as examples. The goal now was to attain emancipation by giving theoretical foundations to technology, where a scientification meant at the same time a consolidation of the institute of technology. Since the 1860s it became customary to appoint outstanding scientists and particularly mathematicians. The attendant problem of more advanced entrance requirements can be mentioned only in passing. There has been a constant controversy between objective requirements and questions of prestige. What settled the debate was the influence of the state's policy of recognition which was directly linked with the struggle for equality of status of the humanist and the non-classical high-school education.

In the first decade after the foundation of the Second Empire the organisational development towards scientific institutes was in a way concluded. The repercussions of the economic crisis of the 1870s had made the importance of technical education particularly topical and brought it to the awareness of the public. It must also be viewed as linked with the way industrial interest groups and the patent movement were organised and becoming political, not least against the background of the important new direction taken by economic policy during these years. By 1880 all German polytechnic schools had officially received the constitution of institutes, although this was not a question of the hoped-for equality with the university. The efforts aimed at this now became particularly prominent, in connection with the now more articulate professional movement of the engineers, whose problems of social self-appraisal and whose standing in society seemed to depend on social and scientific valuation and recognition of their place of education. While polytechnic schools and institutes of technology were often called 'plumbers' academies' or at best regarded as mushrooming specialist schools, which was no less hurtful, engineers were often enough classified as 'higher labourers'. According to their own conviction, as regards social acceptance they had first to assert themselves against the continued resistance of the older academic professions and against the social attitudes fashioned by the educated middle classes, in order to overcome not only prejudices fed on anti-technical criticism of culture but also existing administrative practices of the state. No doubt the social estimation of engineers could not be compared with that of lawyers, medical men or philologians. Schloemilch, a leading mathematician and engineer at the polytechnic school of Dresden, described these conditions in a report to the Saxon government in 1868 (19): "In the leading educated circles of our time the engineer is rated as an upstart and an unjustified intruder into the domain of the truly educated." This sad state of affairs, he went on, could be explained not by ignorance of the mostly disavowed scientific bases of technology, since society, as a rule, understood just as little of the scientific foundations of law or medicine. His observation, that recognition was accorded at best to engineering installations but not to their creators, became a repeated complaint in the efforts at advancement that were regarded and described as a 'struggle of the engineer for recognition by state and society'.

Towards the end of the century this movement, at times, assumed the features of agitation. Since the 1870s it was repeatedly pointed out with increasing national emphasis that in England and France, where supposedly there were no barriers to social integration, engineers stood in incomparably higher esteem. A particular example given was the role of the engineer in the public life of France, where they were to be found amongst the highest civil servants, leading politicians and in the government, while in Germany the representation of technology in the various parliaments was "as good as zero", to quote George Biedenkapp (20), and the engineer in the official organisation of the state's technical services could at best attain the middle grades and was certainly always ranked below the lawyers. The constant polemic against 'assessorism' and lawyers' monopoly reached its climax in the 1880s, in the demand for engineers to be admitted to the higher administrative service and for the training of administrative engineers. All this was often equated with respect and esteem for industry as such. Werner Siemens, an honorary member of the VDI, laid special stress on such arguments in 1876, in his well-known memorandum to Bismarck on the patent system. He stated that "with us, as against other West European industrial countries, there is no place for the engineer", describing the absence of an imperial law to regulate the patent system as a sign of the disrespect for the work and industrial interests of engineers (21). In the discussion that preceded the imperial patent law of 1877 and touched a wider public as well, inventors and engineers, following the socially influential value system underlying the prevailing notion of education, were regarded as mere executors of an immanent technical progress that occurred of itself; therefore they could not rely on a principle of 'intellectual ownership' like artists and writers. Indeed, when the patent law was finally enacted it was not on the grounds of 'intellectual property', a juridical construction of legal theory, but solely on the grounds of economic and national political arguments (22). Mere inventors and discoverers in a trade, as the great Jakob Burckhardt had written in a different context, could nearly always be interchanged and replaced, because they did not concern themselves with the world as a whole, in contrast with the writer, artist and true scientist (23).

For the official estimation and social and scientific valuation of engineering education in Prussia, the transition of the institutes of technology into the domain of the ministry of education in 1879 was a most important event. Engineers felt it as discriminatory and an "incredible and shameful situation", as the VDI had said even in the 1860s and repeated with increasing emphasis after 1870 in resolutions and applications, that in Prussia the affairs of institutes of technology came under the ministry of commerce (which could be explained in terms of its origins), and not (in spite of the basic common cultural interest) like the universities under the ministry of education (24). In the end the question was whether the institutes of technology were to be regarded as purely specialized training institutions or as generally scientific institutes and hence whether they were to be classified with the universities. This question, together with the rejection of an application from the institutes demanding the conferring of the title of professor and the official position of ordinary professors which would give the latter the same service entitlements as were enjoyed by university professors, provided the occasion for the Minister of Education, Falk, to clarify the position in principle. He denied that technology could be scientific, and that the training of engineers in the technical fields was scientific; both on points of principle, arguing wholly in terms of idealist and new humanist notions that 'mere specialist training' was incompatible with 'education through science', and opposing science pursued for its own sake on the one hand to technology as a collection of mechanical rules of procedure and purely practical application of certain scientific results on the other, regarding the two as different in principle. This view indeed continued for some time to mark the prevailing attitudes. Even by the turn of the century one could still read in legal commentaries on the administrative law and conditions of service for teachers at institutes that technology was not science but "art, as applied to the practical needs of life", so that it was not covered by the constitutional freedoms of scientific research and that, in contrast with representatives of the sciences, professors in technical fields merely had the task of teaching and training (25). If, nevertheless, the institutes of technology in Prussia went over into the sphere of education, this was less because of a change in attitude at the ministry than because after Falk was dismissed, the Ministry of Commerce was divided and reorganised in 1879 according to Bismarck's plans. A far-reaching equalisation in status of professors in institutes of technology with their colleagues in the universities was achieved only in 1892, at the urgent request of Emperor William II.

Just as engineers remained convinced that they lacked social recognition, so they themselves insisted on greater general education, basically as a recipe for what one can do oneself to overcome this situation; accordingly there was a demand for expanding the institutes of technology in the so-called educational subjects, namely philosophy, art and literature, history. Along with this went a demand for a stronger *esprit de corps* amongst engineers. Redtenbacher was not only an early expert in his field but one of the first important German engineers who looked beyond the narrow confines of their subjects. Already in 1840, he had commented as follows on the causes of this lower prestige: "If society calls the present state of people in industry crude, it happens to be right." Moreover he had stressed that the concern of the polytechnic school must be not only the development of the technical sciences but, as he put it, the "culture of the industrial public" as such (26), and he himself had made sure that the educational subjects were introduced at his institution. Thus engineers would have to take a greater part in precisely those socially honoured educational products in whose name they felt themselves despised and undervalued, so that they might overcome the scant esteem of the educated classes that formed public opinion. Since the middle classes tried to find their unity above all in the intellectual domain and saw education as the chief instrument of their emancipation, the struggle for advancement and social recognition must be conducted not least in this field. "We declare that German engineers have the same needs for general education and wish to be judged by the same standards as those in other professions with a higher scientific education": these are the words of a corresponding resolution of the VDI which was put forward time and again.

The position of entrepreneurs and industry itself on this question of education could not be defined in a simple manner. In his enquiry on the administration of enterprises and their employees using the example of the firm of Siemens (27). Jürgen Kocka quotes statements made in 1880 by Karl Siemens, then director of the London branch. In a letter to his brother Werner he complained about the lack of general education in his technical staff. As a result, social rejection had adversely affected business in England, especially in dealings with civil servants. He therefore suggested that educated people should be hired, "gentlemen, in order to get away from mechanics who have served their way up". Although, especially in larger concerns, there were thus obvious social expectations and attitudes aiming at a higher general education of engineers, Werner Siemens' reply was nevertheless typical of the ambivalent conception of industry on this point, for he said: "The special skills that we need are fairly independent of education." In the narrow sense, specialised technical skills naturally remained the more important criterion for the needs of industry. Even those empirical technicians who had less of a school education could, on the basis of personal qualifications, rise to positions in the works hierarchy that were likewise held by academically qualified engineers. The state machinery for entitlement which set social scales with formal criteria of education and admission naturally could not be taken over by industry. Here too the bounds for the aims of engineers had to become explicit and their social position determined exclusively by academic examinations and titles. In the competitive situation of industry the individual's productive abilities and the productive knowledge gained outside educational institutions could still succeed even if academic scientific qualifications became steadily more important.

The professional associations of engineers, and the VDI in particular, found it correspondingly difficult to make certain academic training categories into prerequisites for acquiring membership. At the same time this shows what obstacles and problems were bound to stand in the way of introducing a legal protection of title and profession for the designation 'engineer', although people demanded such regulation time and again. The composition of the VDI remained accordingly heterogeneous. Its early constitution defined possible members as practising technologists, teachers of technical sciences and owners of technical establishments. However, it was not only the question of education that was important for defining social position, by other criteria, too, engineers were in the main disqualified. In their steadily growing numbers they were neither independent nor employees of the state, even if within an enterprise their status might suggest to them a comparison with civil servants; but most of them were from the outset in a dependent relation as employees with very different endowments. That is why it was rather difficult to compare them with traditional academic professions, although on the other hand they merely saw the possibility and emphasized the need to become integrated within the ambit of these academic professions and to identify themselves with the corresponding social position. From the observation that engineers could not develop a social self-awareness on the basis of their special training, some sociologists have derived the thesis that they must therefore have adopted the standards of their self-estimation from other social groups. In this connection Bahrdt and later Hortleder have spoken of the lent or foreshortened consciousness of technical intelligence, and made corresponding inferences from it, for example as to what would be its political and social behaviour (28). The historian will be bound to point out that this is a largely untilled field in social history and that in many ways we are more knowledgeable and better informed about the peasant of the fifteenth and sixteenth centuries than about the engineer of the nineteenth century.

The rapid expansion of industry in the period of high industrialisation since the 1880s and 1890s intensified the weighty discussion and debate of problems of engineering training at institutes of technology in which the tension between science and technology was reflected. Relations with industrial practice involved engineers and institutes in new ways all the time, revealing subject matters and methods as incomplete and placing the institutes into the awkward position of having to keep up with technical progress at a pace dictated by industry alone; and this created the impression that with their theoretical starting points the institutes were always trailing behind what industry had already carried out in practice. In the early 1890s all this led to a lasting anxiety of development in the institutes and, before the turn of the century, was to result in thorough changes in their programme of teaching and scientific activity, and also in their external academic position. Above all, a growing specialisation and subdivision of subjects became evident. Until about 1870 there were some 50 to 60 technical and scientific teaching subjects in institutes of technology, while by 1880 it was more than 100, by 1890 nearly 200 and by 1900 at the Berlin Institute of Technology more than 350; and this posed the problem of the scientific nature of technology in a more urgent form than ever before in its history. Released by the critical requirements and the new needs of industry, where a start had long since been made by firms setting up their own laboratories and conducting their own directly practice-oriented research, important changes resulted in technical and scientific attitudes and in the relations between science and technology.

For a long time German industry had been marked by the adoption of West European technical standards. From then on the structure of invention and innovation began to change. The work of invention and development that soon gained in importance in many large enterprises became a division of labour, a process conducted systematically and before long with considerable resources. In the institutes there arose a more strongly observed gap between the foundational mathematical and scientific subjects which, as the critics put it, were oriented by a 'university manner' of approach, and the demands of the technical disciplines required by industry and more strongly pressing for independent practical connections, while displacing the hitherto predominant theorising tendencies. We cannot here discuss the consequent debates of principle which were remarkable not only as regards the scientific history of the technical subjects. What was important is that engineers protested more consciously against the notion that the scientific character of their work resided only in the degree to which it involved the application of mathematics and natural science. The engineer would become scientifically bankrupt, so it was argued, if 'scientific' merely meant 'mathematical' or one-sidedly 'mathematical-scientific'; and the war-cry now was "institutes of technology for the technologists".

In the teaching activities of the institutes, which were officially accorded no more than a training purpose, had until then confined itself essentially to theoretical lectures, construction exercises at the drawing board and demonstration lectures with models. As in the universities, only chemistry students had the opportunity for independent work and experiments. A separate form of experimental research in technology was still fairly undeveloped, and experimental instruction or active and realistic exercises were very unusual. Only now, starting from the indicated developments and demands of industry, teachers in institutes began on an academic basis to draw the wider conclusions from the insight that in the technical fields there was a need for specific experimental teaching and research, but that this required the development of special methods different from those of the natural sciences, namely systematic experiments and measurements on machines and materials in natural measure and under a variety of conditions that corresponded to the real conduct of industrial practice, and that for this one needed special laboratories, devices for measuring, experimenting and testing and general apparatus on a large scale. It was precisely with the corresponding installations in industry in mind that people now declared that institutes of technology had to take over the leadership in engineering research or cease to be called institutes. Within the framework of changes that will here be only hinted at, the German institutes of technology since the mid-1890s developed large scale facilities for teaching and research, corresponding to economically based decisions on scientific policy arising from extensive arguments as to the needs of industry in its competitive situation. Only this gave the institute of technology its modern shape.

The task and position of the institutes were being evaluated from their results with increasing political emphasis, at a time when the industrial sector of the economy began finally to gain the upper hand and German policy turned to the movement of European imperialism. This did indeed mean an important mark in the history of institutes and technology: it was the climax of the developments since Prechtl, Redtenbacher and Karmarsch. Only in this setting were the systematic structure and methods of the technical subjects and of engineering construction further thought through on new foundations and developed by informative research, with the result that the rational and self-contained scientific structure of at least an increasing part of technical activity stood out more clearly; and only this led to the development of the technical sciences as we understand them today. The way leading there, as made intelligible within the scope of engineering training, reveals itself as a process full of change in which the active forces were by no means confined to the often praised dynamic of the field itself from points of view internal to science and technology.

With the conscious grasp of their specific task in research, the institutes of technology at the same time attained a principal criterion for their goal of full equality with the universities, which was a constitutive element of their normative self-understanding, namely the proper prerequisites and conditions for the right to confer degrees which had been so eagerly sought (29). In the last years before the turn of the century, it had been explained by both sides in terms of their own vital interests: for the institutes of technology to attain this right, for the universities to prevent them. After disputes whose acrimony is hard to understand in retrospect, the Prussian institutes and therefore all German institutes were given this right in 1899, as is well known; and this occurred against the bitter opposition of the universities, in the end because of the personal intervention of the Emperor for what he saw as reasons of economic and social policy, more than half a century after the demands aiming at this right had first been made. Even if at first only relatively few could reach degree level, namely those who were then often called the "general staff officers of German industry", nevertheless this question touched very considerably on the interests of all German engineers, for what was at stake was the "long and sadly missed" social recognition of engineers, the "full value of their work even when measured by the standards of traditional learned studies", as it was now put by Adolf Slaby, professor of electrical engineering and personal friend of the Emperor, and an important spokesman for engineers. This was precisely what William II wanted to honour, given his known predilection for technology and its institutes. In this connection he furthermore gave engineers a social task in what he saw as the position of mediators for the workers who were subject to social democratic influences; according to him "the other, humanist orientation has unfortunately been a total failure as regards social matters", which in turn corresponded to the position he adopted in imperial school conferences on the question of equality of status for the real (non-classical) gymnasiums and the humanist gymnasium.

One could not say that the harsh disputes on equality of status for university and institute of technology at the end of the century were conducted on a particularly high level (their acerbity was certainly a peculiarly German matter), a stricture which does, however, apply to both sides. The universities argued with great logical force without change or reservation in the spirit of the antinomies we have mentioned several times before, which had been advanced in the struggle for the social and scientific position of the technical institutions ever since their foundation. What needs noticing here is rather that, as previously stressed, even the sciences represented in them overwhelmingly fitted in with the rest. While people in the institutes of technology renewed with growing self-confidence the reproach that universities were socially blind (a criticism already common before the revolution of 1848 in connection with their attitude towards technology), they praised their own achievements as 'national' ones and engineers as 'pioneers of German value and culture'. Indeed, since the foundation of the Second Empire the institutes of technology had known an unmistakable growth in prestige in all these matters. Constant complaints of social undervaluation were here combined with the argument that engineers should aim at social dominance, a view reminiscent of later technocratic conceptions. More important, and of greater consequences than the acute direct external contrast, was the fact that as regards the state's policies for science, in Prussia and Germany inseparably linked with the name of Friedrich Althoff in those years, a strategy of support for science and research was systematically planned and carried out. This was later to raise the domains of universities and institutes of technology on the one hand and industry on the other onto a higher and necessarily common plane, given that education and science were bound to develop into an increasingly important factor of production.

With equalisation of rank and the right to confer degrees the Dr. Ing. had arrived within the professional group of engineers (as a sop to the universities it had to be written in Gothic script at that time). Above all, however, the newly introduced institute examinations created the likewise new and licensed Dipl.-Ing. Alongside the VDI there soon arose the VDDI, the association of German diploma engineers, within the ambit of technical subject and professional organisations, and these continued to multiply after 1870. The figure of the engineer at the end of the century had not become more uniform as regards levels of education and training. In 1908, Ludwig Brinkmann described the resulting difficulty of fixing the boundaries of their professions, in saying that "in a modern technical enterprise of medium size we often find men of supreme scientific education with academic degrees working on the same job with the same rights as self-taught men who have never been to a specialist school" (30). For the vast majority of engineers their economic and social position was, in fact, largely determined by the problems of the new group of employees. Their social self-esteem remained insecure. According to the results of an investigation sponsored by the VDI (31), even in 1959 more than half its members still saw themselves as undervalued as regards their professional qualification, still finding the reasons for this in an inadequate measure of socially significant general education, and in a lack of feeling of solidarity. Thus, they saw themselves as ever caught in the traditional antinomy of education and training; and in spite of their academised status, in that "awkward world position" which, as mentioned at the beginning, Max Maria von Weber had diagnosed a full century before.

Notes

1. R. von Mohl, *Die Polizeiwissenschaft nach den Grundsätzen des Rechtsstaates*, Part 1,2.A, Tübingen, 1844.
2. *Staatslexikon*, Part 4, 1837, pp. 676ff.
3. Cf. W. Seibicke, *Technik. Versuch einer Geschichte der Wortfamilie in Deutschland vom 16, Jahrhundert bis etwa 1830*, 1968.
4. H. Schimank, 'Das Wort "Ingenieur", Abkunft und Begriffswandel', in *Zs. d. VDI*, Part 83, 1939, 325ff.
5. Cf. École Polytechnique, *Livre du Centenaire*, Paris, 1898.
6. M. M. von Weber, 'Über Bildung der Techniker und deren Prüfung für den öffentlichen Dienst', in *Der Ingenieur*, Zs. f. d. gesamte Ingenieurwesen, 1854, pp. 99ff.
7. *Zs. d. VDI*, 1857, 1ff.
8. Cf. for this K. H. Manegold, Universität, Technische Hochschule und Industrie, Ein Beitrag zur Emanzipation der Technik im 19. Jahrhundert, 1970.
9. Cf. C. F. Nebenius, 'Über technische Lehranstalten in ihrem Zusammenhang mit dem gesamten Unterrichtswesen und mit besonderer Rücksicht auf die Polytechnische Schule', Karlsruhe, 1833.
10. P. Chr. W. Beuth, 'Glasgow', in *Verhandlungen des Vereins zur Beförderung des Gewerbfleißes in Preußen*, Part 3, 1824, pp. 169.
11. F. and P. Goldschmidt, *Das Leben des Staasrates Kunth*, 1881, 269ff.
12. Cf. J. Neuwirth, *Die k. k. Technische Hochschule in Wien 1815-1915*, Wien, 1915.
13. F. C. Biedermann, 'Wissenschaft und Universität in ihrer Stellung zu den praktischen Bedürfnissen der Gegenwart', Leipzig, 1839.
14. Cf. Fr. Schödler, 'Die höheren technischen Schulen nach ihrer Idee und Bedeutung', Braunschweig, 1847.
15. Peter Lundgreen, 'Bildung und Wirtschaftswachstum im Industrialisierungsprozess des 19. Jahrhunderts, 1973'; ders., Techniker in Preußen während der frühen Industrialisierung, 1975.
16. Cf. F. Grashof, *Redtenbachers Wirken zur wissenschaftlichen Ausbildung des Maschinenbaues*, 1866.
17. Cf. K. H. Manegold, 'Die Entwicklung der Technischen Hochschule Hannover zur wissenschaftlichen Hochschule', in W. Treue (ed.), *Naturwissenschaft, Technik und Wirtschaft im 19. Jahrhundert*, Part 1, 1975, pp. 185ff.
18. F. Reuleaux, *Lehrbuch der Kinematik*, 1875ff.

19. Cf. *125 Jahre Technische Hochschule Dresden*, 1953, p. 41.
20. *Der Ingenieur, Seine kulturelle, gesellschaftliche und soziale Bedeutung*, 1910, p. 25.
21. Cf. K. H. Manegold, pp. 79ff., Note 8.
22. Cf. K. H. Manegold, 'Der Wiener Patentschutzkongress von 1873', in *Technikgeschichte*, Part 38, 1971, pp. 158ff; *ibid.*, 'Vom Erfindungsprivileg zum "Schutz der nationalen Arbeit"', in *Zeitschrift der Technischen Hochschule Hannover*, No. 2, 1975, 12ff.
23. In his 'Weltgeschichtlichen Betrachtungen' (here: Berlin 1960, p. 183).
24. Cf. for this, K. H. Manegold, pp. 70ff., Note 8.
25. Cf., e.g., C. Bornhack, 'Die Rechtsverhältnisse der Hochschullehrer in Preußen', 1910.
26. Cf. R. Redtenbacher, 'Erinnerungsschrift an Friedrich Redtenbacher', 1879, p. 33.
27. 1969, p. 169f.
28. Cf. G. Hortleder, 'Das Gesellschaftsbild des Ingenieurs. Zum politischen Verhalten der technischen Intelligenz in Deutschland', 1970.
29. Cf. K. H. Manegold, 'Technische Forschung und Promotionsrecht', in *Technikgeschichte*, Part 36, 1969, pp. 291ff.
30. In *Der Ingenieur*, 1906.
31. 'Der deutsche Ingenieur in Beruf und Gesellschaft', *VDI Information*, No. 5, 1959.

THE COMING OF THE ASSEMBLY LINE TO EUROPE

PATRICK FRIDENSON
Université Paris X — Nanterre

Mechanization can be analyzed from several view points. A history of technology *per se* will concentrate on the long sequel of inventions which have led to the present tools and objects. A cultural approach will "describe the impact of a mechanized world on the human organism and on human feelings" (1). The economic historian will show that a technological dynamic is inseparable from 'a managerial dynamic' and 'a personnel dynamic', which are the three basic elements in the modern factory system (2). Anthropological history will deal with the adaptation of technology according to various national needs. Our purpose here rather fits in the frame of social history. We want to see what exactly are the new relations between the agents of production that become possible with the use of machinery.

Therefore, our attention was attracted by the labour-intensive industries. In that area the symbol of the twentieth century factory is the moving assembly line (which, in fact, concerns about 25% of the labour force). It established the pattern of manufacture and of consumption which exists today. This paper limits itself to a case study of this important innovation, i.e., its transfer to Europe. It is clearly no more than a small step in research, because primary sources are sparse and detailed treatments of so wide a topic are still rare. It falls into three parts. First, it tries to determine a chronology. Then it discusses the European managers' motivations for the diffusion of Henry Ford's technique. Finally it examines the workers' resistance to the new labour process (3).

1. Chronology of a Diffusion

The transfer of the assembly line to Europe may have been made easier by some European precursors' attempts at continuous production.

159

Krohn/Layton/Weingart (eds.), The Dynamics of Science and Technology.
Sociology of the Sciences, Volume II, 1978. 159–175. *All Rights Reserved.*
Copyright © 1978 by D. Reidel Publishing Company, Dordrecht, Holland.

The emphasis has been laid on the early beginnings and quiet growth of flow production in the United States. But Europe, though more slowly, also made some steps toward the assembly line. Indeed, "all the 'components' of flow-line production had been used with success . . . in several countries" (4). Britain and France experienced significant developments in material handling. In England, the manufacture of ships' biscuits used, first, a human assembly line (1804), then a mechanical process (1833). In the 1830s, material flow was a subject of interest for British engineers, such as Nasmyth, whose standardized machine tools heralded the automatic tools and whose factory pioneered the use of the assembly line, and Bodmer, who constructed some of the first travelling cranes and endless belts (5). In France, also, attempts were soon made to solve the problem of conveyance within the production. Among these, we find the conveyors used in building the Suez Canal in the 1860s, and the conveyors working in some roof tile factories which were operating during the same period (6). As for interchangeable-parts manufacturing, it had its origins in Europe, first in Sweden, with Polhem, and in France, with Blanc (7). After having been strongly developed in America, it came back to Europe in 1858, when the British small-arms industry adopted this practice. Thus we may say that the two manufacturing traditions which Henry Ford brought together were present in Europe, too.

Even if we come to the pre-war era when Ford's engineers introduced the moving assembly-line in the automobile production, some European car makers were not too far away. In England, in 1913, the Sunbeam factory at Wolverhampton assembled chassis by moving them mechanically from one group of specialist fitters to the next. It was on the verge of moving assembly. In Italy, in 1913 too, the large Fiat works of Turin were developing flow assembly. In France, after 1907, the Lorraine-Dietrich factory at Argenteuil set up a motionless line. In some of the Renault shops at Billancourt, and especially in the Berliet works at Lyons, there was much mechanization of materials handling and assembly. However, these European firms did not contemplate a quick move to mass production, because of their wide variety of products or of their small amount of sales (8).

Another interesting early example of highly mechanized production in Europe can be found in another new industry. The Gnome aircraft engine factory at Paris, in 1913, relied heavily on the use of automatic lathes (9).

It was thus the First World War which urged (and helped) the Europeans to follow the trend epitomized by Henry Ford and his engineers in 1913–1914. The war created a mass demand, for various types of armaments and munitions. The entrepreneurs had to satisfy it by all means. Some of them decided to adopt the American-born assembly line (10).

Two European motor car firms started doing so in 1915. In the Walthamstow factory of the Associated Equipment Company a moving-track assembly line was installed. The contemporary press enthused warmly over the technique which produced a 3–4 ton lorry every half-hour of the day. This highly efficient method was virtually unique in the country at the time (11). At the Renault plant in Billancourt, moving conveyors appear in some shops which are geared to making parts for aircraft engines. They carry the parts along four lathes at a speed of 1.35 m min^{-1} (12). 1916 was Italy's year. Giovanni Agnelli launched the Fiat factory at Lingotto. There he decided to renovate his methods and began a program, including assembly lines, in a few departments (13). In 1917 it was Marius Berliet's turn, when he too built a new factory, at Vénissieux. It contained two assembly lines, "just like Ford's", to produce lorries (14). The same year, at Renault, the production of the tank was another step forward, as it was carefully broken into various stages and relied on standardized parts. The assembly hall (in 1918) was crossed by double rail tracks on which small trolleys moved about, each of them carrying a chassis (15). Parallel progress was made in the production of shells. It was most integrated at the major French shellmaker's plant. Citroën's, in Paris; it was also a brand new one, built in 1915 (16). But even a smaller shell producer, Delage (again a former automobile firm), which had a large new factory at Courbevoie since 1911–1912, installed assembly line work (17).

From these examples the automotive industry appears to have played an important part in the transfer of the assembly line to Europe. How important cannot be exactly said until manufacturing procedures in other industries are investigated. We know of only one application elsewhere. By 1916 the French firm l'Eclairage Electrique, in Suresnes, was operating a line in continuous movement, which mechanically carried the components past successive groups of workers (18).

The war had a two-fold effect. It probably accelerated the adoption of the line by the Europeans, who had to answer quickly the challenges of mass production and manpower shortage. In this case the innovation was transmitted by a process of stimulus diffusion before any direct diffusion could occur (19). But at the same time, the war may have delayed its spreading, inside the factories, from one sector of the economy to another, from one nation to its neighbours. Indeed, the war gave businessmen opportunities to issue a large variety of products, and not to concentrate on one item. Others would think that to convert an entire factory to flow production would take more than one year. Was it worth-while when the needs of the armies were so quick to change and nobody could tell how long the war would last? As for the German manufacturers, they deliberately refused to increase the deskilling of their labour force because of socio-political circumstances (20).

When peace came back, the question became much clearer for European industrialists. Books and newspapers heralded the success of the American experience. They invited the more progressive firms to copy it (21). Managers had to appreciate whether a mass market for their products could be developed; if so, could they face the high capital requirements for the new production facilities? Had they the engineering talent available? What was the situation of the labour force? The answer varied. Therefore the assembly line diffused at an uneven pace. Each country, each industrial sector would have its own rhythm and style.

Let us consider, for instance, the European motor vehicle industry. All the lines were built after some engineers had been sent to the United States to study the advanced production methods of which their managers had read the published accounts. In some cases, American engineers were hired and brought to Europe: as did Citroën in 1923 (22). Still the lines slightly diverged from the American originals (23). They were specifically designed to fit the European needs. In the Renault works in 1922, each car stayed at a station for 40 minutes, a rate of assembly which was quite slow by contemporary American standards (24). Even the European subsidiaries of the American manufacturers had to adapt to the national needs and constraints. Evidence of this was given by Henry Ford's British factory: "Unlike the Highland Park plant in America, Manchester did not have the famous 'body drop', bringing coachwork down from above to avoid the problem of lifting it on the chassis, as would be the case here" (25). The gradual way in which the new methods were introduced in the auto factories should also be emphasized. So as to maintain current production without too much confusion, procedures tended to be changed department by department. Also, the potential of skilled engineers which was at the firm's disposal limited the speed of diffusion.

Anyway, France took the lead. Citroën started an assembly line for its first postwar model in – or after – 1919, and so did Berliet from 1920 onward. Renault followed in 1922, and Peugeot a little later (26). The other countries were much more cautious. Germany waited until 1924, when only the largest car-maker, Opel, began to instal the conveyor system (27). It reached other firms much later (28). The main national car-maker played the same introductory part in Italy, with Fiat spreading the line in its works behind Citroën, and in Czecho-Slovakia, with Skoda building in 1925 only its new car factory significantly called 'America' (29). The British situation is much more diversified. The leading firm, Morris, contented itself with stationary line production after 1919. Groups of workers stayed at stations on the line, and the chassis was simply pushed by hand from one station to another. William Morris was very reluctant to invest in manufacturing. But Morris's greatest rival, Austin, introduced the moving assembly line system between 1922 and 1925 in his Longbridge plant, which was more highly-integrated and more sophisticated than Morris's Cowley, as Austin was always keen on efficient production methods (30). Nevertheless, they both seem to have lagged behind Ford-England, which is said to have used a line after 1920 both in its main plant at Manchester and its new branch factory at Cork, in Ireland (31). Only in 1934 was the moving assembly line installed in Cowley, as the Morris company rebuilt the whole factory. The conveyors and all the necessary machinery were bought from the firm of Bagshawe, located at Dunstable (Bedfordshire). They probably used American patents, as often was the case when American machinery itself was not acquired (32).

Altogether, by 1930, the moving assembly line had made its break-through in the largest European motor car firms. But it also succeeded in some other industries. Such were electrical engineering and clockmaking. Assembly lines are to be found in the German electrical engineering works of the late 1920s. The french firm Lip in Besançon became the nation's leader in that area by mass producing watches (33). In 1927 the French Northern Railway Company introduced the line in its railroad car shops.

So much for the historical evidence of the European adoption of the moving assembly line. It is time now to assess its economic and social meaning: why was it promoted by entrepreneurs in the twenties?

2. The Extension of Managerial Control

Since the 1890s, European manufacturers and engineers had been gradually modifying the traditional methods of shop management. Modern industrial administration emerged, particularly with Taylorism. Taylorism was already propagated in Europe before 1914. It developed its influence during the war and in the twenties. Therefore, the central question is what specific reasons determined the (relative) success of the assembly line: economic? social? ideological? Compared with Taylorism, why would European business leaders care for it?

What Taylorism offered within the plant was the setting of new norms of work (34). But what was novel about Ford's innovation? European managers at once said it was the application of machinery to a specific part of the production process; in other words its advanced methods of materials handling and inplant transportation (35). They could reap two major benefits from their introduction.

First, the line allows the management to reduce the production time to a minimum, and to limit strictly the amount of fixed assets. Both French and English experts of the interwar period praised the line for "the elimination of waste periods in the day's work" (36). Thanks to the overhead cranes, conveyors and transveyors, the individual parts move regularly past each worker. 'Ever Onward' was the motto at the Cowley car factory in 1934. There "ingenious endless chain conveyors . . . travel continuously round the shop in order that every group of workmen shall have at hand the components they require, and yet be unhampered by supplies" (37). Consequently, three economic advantages arise, which were stressed by the Renault engineers as early as 1918. This is a labour-saving device, since a much smaller proportion of labourers is needed. This is also a space-saving procedure, because it gives more floor-room than the former plans of organization. And it makes possible a better, closer control over the production flow. Hence, a larger output together with lower management costs (38). To sum up, the substitution of machinery for labour improves the circulation of capital. It is conveniently illustrated by the image of the mechanical hand which a British journalist applied to the new Cowley plant: "An ultra-modern conveyor system, as that in operation at Cowley, is really a mechanical hand with, literally, hundreds of fingers, each carrying and handing out appropriate parts, at appropriate periods to appropriate places" (39).

Second, with the moving assembly line, manufacturers can impose on the workers a higher intensity in the performance of tasks. While parts and materials go on moving, the worker, who hitherto used to walk about for tools and materials, will now remain for ever at the same station (40). To quote again descriptions of the Cowley factory, "things are brought mechanically not only to him, but right to the exact spot on the chassis needing them" (41). This means that the worker is deprived of the very small degree of control over the content, speed, rhythm of work which had been left to him by Taylorism. Losing the last remnants of autonomy and initiative, he functions as an appendage to a machine. "Synchronised conveyors have brought up all the parts, all the time to all the right spots at the exact moment each is neeeded" (42). Subordination is increased, because the worker is paced by the line. The labour process is designed around the performance of the machine, and the worker has to perform in accordance with its needs. His output is determined mechanically by the speed given to the conveyor which goes near him. The latter is "so timed that every man has just a little more time for his job than is strictly necessary" (43). And while the line itself relies on a system of ever more complex machines, the tasks required from the workers are fragmented into ever more simple operations. This minimizes dependence on specific and often scarce labour skills. As a French engineer put it: "Skilled workers demand high wages. So they cannot be committed any more to assembly work . . . . The new methods care very little for the worker's technical ingenuity . . . . Thus, most skilled workers have been eliminated. From now on each worker always does the same job, and . . . it is easy for him to reach the dexterity" which becomes the chief requirement for successful operation. "He is now able to work almost by reflex, auto-matically, and to the maximum extent of his powers" (44). Thus, the assembly line makes labour cheaper and more easily replaceable. It has still another advantage. When the shops are set in line, the foreman can keep a sharper eye on the workers (45). Further on, the tasks are standard-ised, routinised and predictable because management has also provided for the standardisation of machines, tools and parts.

The essence of Taylorism had been labour efficiency. But those managers who developed the assembly line in Europe wanted something more: efficiency in capital spending. They found it with the new norms of production which Ford's innovation made possible.

But in the United States, the contributions of Ford were not confined to the productive process. They included also "the enlargement of a mass market by low prices and high wages" (46) or, to put it otherwise, Ford linked the assembly line with another innovation, the establishment of social norms of consumption (47). The European experience showed some striking differences. European managers shared Ford's original concern with lower selling prices, and during the 1920s they achieved an important drop in real prices, which was noticed by many contemporary observers, especially in the British and the French automobile industry (48). But the rise in the workers' wages was much smaller than in America.

Nothing here was worthy to be compared with Ford's $5 per day wage, which nearly doubled the going rate of pay for American auto workers. European managements conceded to labour a smaller share of the productivity gains resulting from the assembly line. For instance, at the Renault works, in 1925–1926, a 22% wage increase was granted to those working in the pushrod shop after the introduction of the line. Under the same circumstances, assemblers of 6 hp engines got a 12% pay increase (49). Employers and economists explained that, contrary to American firms, the European ones were very dependent on foreign markets and had to overcome their protection by tariff barriers. The strong competition abroad set limits to the rise in wages (50). A few other reasons can be added. The strength of the labour movement and its bargaining power were declining in the twenties. Except for a few pioneers like Citroën (51), European managers were not really interested in workers' mass consumption. The new market they aimed at was still the lower middle class. Because of their policy and, moreover, of the lower average income earned by the Europeans, the output of mass-produced goods rose less quickly in the 1920s than in the U.S. of the 1910s.

The introduction of the assembly line could also bring some benefits to the employers in the field of labour relations. The Russian writer Ilya Ehrenburg gave his version of this argument in a 1929 chronicle about the Citroën works: "You can't lower the cost of materials, but you *can* lower the cost of labour. The year 1919 was past. The workers' committees were disbanded. The strikes had failed. Monsieur Citroën showed his workers a new toy from overseas: The belt, the moving conveyor belt. Let the workers groan, their grumbles would be drowned out by the noise of the new presses. Citroëns were cheaper than ever" (52). In fact, the line could serve the business community in two ways. This was the time when, all over Europe, the working day was reduced to eight hours, and managers were preoccupied by the desire to counterbalance this change by a higher labour productivity. The assembly line was certainly a very adequate instrument for this purpose (53). The line was also the best means for employers to recruit and use a new labour force. For the unskilled and semi-skilled workers were often migrants, of rural or foreign origin, and women. In the Citroën factories of 1929, "people came from different places. There were Parisians and Arabs, Russians and Bretons, Provençals and Chinese, Spaniards and Poles, Africans and Annamites. . . . Now they were all at the same conveyor belt" (54). In post-1926 Britain the motor car industry "drew on the vast supplies of destitute people in the depressed regions". A large proportion of this labour came "from the ranks of the agricultural labourers" and also "from the distributive trades" (55). This heterogeneous recruitment outgrew the existing divisions of labour. It also weakened the already declining trade unions. Being based on skilled workers, they did not easily accept and incorporate semi-skilled workers.

Naturally, the assembly line was also linked with the cultural and political appeal of Fordism in the twenties. Industrial productivity was "the practical solution to the labour question": it had to be "Ford or Marx", a German commentator said (56). Fordism evoked enthusiastic responses, particularly in Germany (57). For instance, a selected biblio-graphy on scientific management in Germany, which was prepared in

1926, contains a list of 21 different titles on Fordism (58). Fordism also became the vogue in France and in Italy. The managers of the Fiat company heralded the emergence of a "progressive capitalism" which promised abundance and welfare provided "social pacification" and industrial discipline were achieved (59). Whereas divergent tendencies could develop within Taylorism (60), Fordism had a one-dimensional thrust. It "best served the bourgeois-conservative . . . ends of European business and industry in the later 1920s". The American vision of productivity was thus divested of Taylorism's "Utopian implications". Fordism "refurbished the entrepreneur directly" (61).

Therefore we can agree with E. Layton that mass production in Europe became "a means of enhancing the power of the corporation over the workers" (62). Yet it did not raise a coherent opposition from the part of labour.

3. The Conflicting Responses of Labour

The problem then is to analyze the responses of European workers to the Fordist organization of production. A clear-cut distinction should be made between labour's leadership and the rank and file.

The European trade unions had fought hard battles against Taylorism before and during the war. But gradually they moved to the acceptance of Taylorite methods, so long as they did not imply mere speed-up. They committed themselves to the cause of economic growth. Obviously the assembly line could serve the same ends. So they accepted it without much discussion. Such was, of course, the attitude of reformist, non-Marxist unions. The British trade union officialdom continued its policy of output exhortation (63). The French Confédération Générale du Travail saw, with direct and immediate interest, the greater production that the line would bring about. Concrete benefits for the working class could be derived from it, in terms of higher real wages, shorter hours and cheaper manufactured products (64). Conversely, the CGT insisted on the organization of the workers and on the bargaining of higher wages as a reward for increased productivity. It expected also some ideological benefits of Ford's innovation: it could lay the basis for a productivist society where the unions might reach some form of accommodation with business and the State. But what about Marxist unions? Apparently, the Communist-led Confédération Générale du Travail Unitaire was much more hostile to Fordism. It had always faced all types of "Capitalist rationalization" with outright opposition. However, in 1927 the minority group inside the CGTU Metalworkers' Union (it consisted of revolutionary syndicalists and of anarchists) urged their Communist leaders to take a firm stance against the moving assembly line itself. The majority of the union refused. They argued that the line was a convenient means to produce more goods, at a cheaper price and with the same effort. This is the reason why it was used in Soviet Russia, where workers worked only 7 hours a day. The line was a finding of science. You cannot fight against science or machinery more than you can fight against the rain. When the rain falls, the only sensible thing to do is to take an umbrella. Therefore, the union's official policy could only be to struggle over the consequences of mass production. The assembly line should stop for 10 minutes every second hour; its speed should be slowed down; no woman or youth under 21 should be allowed to work on an assembly line. These claims were accompanied by the usual demands for higher wages and shorter hours. After all, that attitude looked rather similar to the CGT's. Moreover, in 1932, the CGTU dropped its detailed demands on the pace of the line, and limited its collective action to wages. Fordism, the union stated, is a mere technical change. It must be retained by a socialist country. Under a capitalist régime its nuisances grow out of capitalist control. "Rationalization must be fought before revolution, and achieved after its victory" (65).

So, Marxist and non-Marxist unions finally drew parallel conclusions. How do we explain this? Certainly by looking at the social structure of both their membership and leadership. As previously mentioned, they rely mostly on skilled workers. This group of workers in those years favoured and admired technical change. They could understand the managerial emphasis on production and effort. But their reaction was not only an ideological one. The line, they believed, would not alter their higher position among factory workers. Having undergone Taylorism, they apparently had little to lose with Fordism. Maybe they had something to earn: indeed they spread to new trades and departments in the factory, such as supervision, maintenance, inspection, toolroom, planning and drawing offices (66). These new jobs brought to some of them "significant short-term improvements in their status and earnings" (67).

A striking exception to this overall dedication to production was the Italian Communist leader, Antonio Gramsci. In the early 1930s he maintained that Fordism was a definitely capitalist technique for the making of a more disciplined, stabilized working class. A socialist power would have to transform its characteristic features (68).

Let us now try and tell the same story 'from the bottom up': How did the rank and file workers react, particularly semi-skilled workers? They often appreciated the comparatively high wages which they could earn on the line (69). They complained about the speed up. But, once we have said that, we must admit that their actual responses varied enormously. In Italy, with its peculiar political conditions, there was "no apparent resistance" (70). Gramsci pointed out that Italian skilled workers had never, individually or collectively, "actively or passively opposed innovations leading towards lowering of costs, rationalization of work or the introduction of more perfect forms of automatism and more perfect technical organization of the complex of the enterprise. On the contrary" (71). In Britain, "methods such as the assembly line were regarded with suspicion and vigourously opposed by the workers" in the engineering industry (72). In France, there was a certain amount of resistance against the line among grass-roots people. Like in the United States, some workers continued an informal resistance: restriction of output, absenteeism, turnover remained usual practices. But one often forgets that a formal resistance also took place, in the shape of strikes. In 1924, workers struck at Citroën when the assembly line was introduced in the sheet-iron works. They did too in 1926, when the polishers came to experience the line. The 1927 strike at Citroën started with grievances against the pace of the line. In 1928 and 1929 strikes broke out after the establishment of assembly line by smaller car-makers, Talbot and Donnet, or by a larger one in a new outlet: Renault at its body-shop of la Plaine Saint-Denis. In 1930, a 2000 workers firm, the SMT in Billancourt, struck when a line was built in the regulation shop (73). So did the Lip workers in Besançon. Several questions may be asked about these conflicts. Who initiated them? In many cases, the typical mass production operators, i.e., labourers and semi-skilled workers — among them migrants and women.

See, for instance, the 1927 Citroën strike. But in other struggles skilled workers shared the initiative. What were their motivations? Industrial discipline, which they rejected, wages which did not answer their expectations, and fatigue (74). What was the Communist unions' position? Some of the strikes were wild-cat strikes, outside a union's responsibility (75). Some other ones were under the control of a union, which immediately focused them on wages. This reveals a growing discrepancy between even revolutionary unions and some rank and file workers. The leadership still criticized the loss of autonomy which resulted from the assembly line (76), and still demanded the institution of shop stewards. But it had slowly abandoned the old goal of actual workers' control at shop and factory level. While the old spirit had found some adepts in the renewed labour force, among organized and unorganized workers, hence generating various forms of resistance. The reader should be reminded that strike was not workers' inescapable reaction: at Cowley in 1934 the coming of the conveyor did not give rise to any strike (77). The reactions might be much more covert. But they persisted to some extent despite the official acceptance of the assembly line by labour organizations.

4. Conclusion

The introduction of the assembly line in Europe was not a mere technological event. Economic, social, cultural needs and resources determined its European diffusion, for which the adaptability of modern technology was responsible.

The various statements previously quoted show convincingly why dynamic European managers were interested in the assembly line. The reorganization of the labour process would permit a quicker movement of capital and thus improve the conditions of capital accumulation. Yet, for various reasons, we have mentioned, the European entrepreneurs developed the assembly line to a smaller extent than their American counterparts. In the climate of the postwar period, the diffusion of the line also played its (small) part in a larger strategy: "recasting bourgeois Europe" (78).

By contrast, the European working classes displayed less coherence. Large numbers of workers did not oppose "to Americanism as such, nor even to its attendant effects in social life, but rather to the specific form it would take in conditions of intensified economic exploitation" and – in Italy – "authoritarian cultural repression" (79). But other workers kept resisting a process which increased their real subordination. Worker' response to mass production is indeed complex: "person-machines relationships are affected by many external variables, such as the level of national industrialization, how management uses skills, how workers integrate their social life in the department, and other factors" (80) such as wage systems. The line did not make employees more antagonistic to industrial jobs. Simply, a certain degree of antagonism remained, till the present outburst of 'blue-collar blues'.

Hence the present attempts at job enlargement. But do they depart from most basic principles of mass production as embodied in the assembly line?

Notes

1. S. Giedion, *Mechanization Takes Command*, New York: Oxford University Press, 1948, p. 124.
2. D. Nelson, *Managers and Workers. Origins of the New Factory System in the United States 1880-1920*, Madison: University of Wisconsin Press, 1975, p. ix.
3. My research in Britain was made possible by a F. W. Deakin fellowship at St. Antony's College, Oxford. I am very grateful to my colleagues J. M. Laux, A. Moutet, D. G. Rhys for the information and the kind advice they gave me on this matter. My thanks also to Mr K. Revis, coordinator of the Historic Division of British Leyland, and to Mr G. Hatry, coordinator of the Section d'Histoire des Usines Renault.
4. Ray Wild, 'The Origins and Development of Flow-Line Production', *Industrial Archeology* 11, 1974, p. 51.
5. Giedion, *op. cit.*, 1948, pp. 87-93. Wild, *op. cit.*, 1974, pp. 49-50.
6. See the drawings kept at the Bibliothèque Nationale, Paris, Cabinet des Estampes.
7. Robert S. Woodbury, 'The legend of Eli Whitney and Interchangeable Parts', *Technology and Culture*, Summer 1960, pp. 242-247.
8. James M. Laux, 'Le modèle européen', in J. P. Bardou, J. J. Chanaron, P. Fridenson, J. M. Laux, *La Révolution Automobile*, Paris: Albin Michel, 1977, p. 91.
9. P. James, 'Une Usine Modèle', *l'Aérophile* 20, 1913, pp. 412-413.
10. Albert Thomas, 'Preface', in P. Devinat, *L'Organisation Scientifique du Travail en Europe*, Geneva: Bureau International du Travail, 1927, p. vi.
11. 'A.E.C. 50 years', *A.E.C. Gazette*, November-December 1962, pp. 7-8.
12. *Bulletin des Usines Renault* 1, 15 December 1918, pp. 13-14.
13. V. Castronovo, *Giovanni Agnelli*, Turin: Unione Tipografico — Editrice Torinese, 1971, pp. 157-159, 328.

14. M. Laferrère, *Lyon Ville Industrielle*, Paris: P.U.F., 1960.
15. Archives of the Section d'Histoire des Usines Renault (Billancourt), note on the Renault works, 1918.
16. André Citroën, 'L'organisation scientifique de l'usinage', *L'information ouvrière et sociale*, **1**, 7 March 1918.
17. A. Petitet, *Organisation rationnelle d'une usine travaillant en séries, et montages d'atelier*, Paris: Dunod, 1920.
18. Archives of the Service Historique de l'Armée (Vincennes), 10 N 59, réponse de la société l'Eclairage Electrique au président de la Commission des Armements et Munitions, October 1916.
19. For these concepts, cf. Edwin Layton, 'The Diffusion of Scientific Management and Mass Production from the U.S. in the Twentieth Century', *Proceedings of the XIV'th. International Congress in the History of Science*, Tokyo, 1974, pp. 377–378.
20. G. Feldman, *Army, Industry and Labor in Germany 1914–1918*, Princeton: Princeton University Press, 1966, p. 206.
21. G. Cote, *L'automobile après la guerre*, Paris, 1918. V. Cambon, *L'industrie organisée suivant les méthodes américaines*, Paris, 1920. O. Prätzel, *Von der Fabrik-Organisation*, Braunschweig: Westermann, 1919. And so on . . .
22. *Boston — Massachussets Transcripts*, 23 April 1923.
23. Layton, *op. cit.*, 1974, p. 381.
24. Pierre Maillard, 'Les procédés modernes de montage des châssis d'automobiles. Le montage à la "chaîne" de la 10 HP Renault', *Omnia*, 3 September 1922, p. 203.
25. M. E. Ware, *Making of the Motor Car 1895–1930*, Harlington: Moorland Publishing Company, 1976, p. 24.
26. C. Rocherand, *L'Histoire d'André Citroën*, Paris: Lajeunesse, 1938, p. 63. *L'Usine*, 1 July 1920. Maillard, *op. cit.*, 1922, pp. 202–206.
27. A. Pound, *The Turning Wheel*, Garden City, N.Y.: Doubleday, 1934, p. 250.
28. H. Scholz, 'Rationalisierung in der Lastkraftwagenindustrie', *Verkehrswesen*, **5**, January 1930, pp. 68–71.
29. Castronovo, *op. cit.*, 1971, p. 328. J. Hausman, 'La technique automobile au coeur de l'Europe', *Ingénieurs de l'Automobile* **45**, 1972, p. 560.
30. P. W. Andrews and E. Brunner, *The Life of Lord Nuffield*, Oxford: Blackwell, 1955, pp. 87–88. L. P. Jarman and R. Barraclough, *The Bullnose and Flatnose Morris*, Newton Abbot: David and Charles, 1976, pp. 82–84. Interview with Mr. John Pringle, 9 March 1977 (Mr. Pringle was a draughtsman at Cowley in the 1920s). Z. E. Lambert and R. J. Wyatt, *Lord Austin the Man*, London: Sidgwick and Jackson, 1968, p. 142.
31. Information supplied by Professor D. G. Rhys. Ware, *op. cit.*, 1976, writes on page 24 that "production line assembly here was also automated" by 1914. This interpretation seems questionable.
32. C. R. Lucato, 'At Cowley Today', *The Morris Owner*, August 1934, pp. 556–558. M. Sampson, 'Cowley Cameos. I. The Conveyor System', *The Morris Owner*, November 1934, pp. 863–864. Andrews and Brunner, *op. cit.*, 1955, p. 197. Letter from F. Dyer (who was with Bagshawe), 25 February 1977. Interview with John Pringle, 9 March 1977.
33. R. Brady, *The Rationalization Movement in German Industry*, Berkeley: University Press, 1933. P. Daclin, *La crise des années 30 à Besançon*, Paris: Les Belles lettres, 1968, p. 27.

34. B. Coriat, *Le taylorisme, le fordisme, la production de masse et les nouveaux modes d'organisation du travail industriel,* Université Paris X — Nanterre: unpublished Ph. D., 1976, pp. 46–114.
35. *Bulletin des Usines Renault* **1**, 15 August 1918, p. 4. 15 September 1918, p. 3.
36. Maillard, *op. cit.,* 1922, p. 202. Lucato, *op. cit.,* 1934, p. 556.
37. Lucato, *op. cit.,* 1934, p. 558.
38. *Bulletin des Usines Renault,* **1**. 15 October 1918, p. 2.
39. Sampson, *op. cit.,* 1934, p. 863.
40. Maillard, *op. cit.,* 1922, p. 203.
41. Sampson, *op. cit.,* 1934, p. 864.
42. *ibid.*
43. Lucato, *op. cit.,* 1934, p. 558.
44. Maillard, *op. cit.,* 1922, p. 202.
45. *Bulletin des Usines Renault,* **1**, 15 October 1918, p. 4.
46. C. S. Maier, 'Between Taylorism and Technocracy: European Ideologies and the Vision of Industrial Productivity in the 1920s', *Journal of Contemporary History,* **5**, April 1970, p. 55.
47. M. Aglietta, *Régulation et crises du capitalisme, L'expérience des Etats-Unis,* Paris: Calmann-Lévy, 1976, pp. 96, 135–138.
48. Maillard, *op. cit.,* 1922, p. 202. Lucato, *op. cit.,* 1934, p. 556.
49. Private papers of Mr. Jean Guittard, original tables of 1926.
50. B. Birnbaum, *Organisation der Rationalisierung Amerika-Deutschland,* Berlin, 1927, pp. 70–71. National Archives, Paris, F 22 185, letter from the Renault management to all the workers, 21 May 1926.
51. A. Citroën, 'L'avenir de la construction automobile', *Revue politique et parlementaire,* 10 May 1929, p. 249.
52. I. Ehrenburg, *The Life of the Automobile,* New York: Urizen Books, 1976, p. 27.
53. National Archives, Paris, letter from the Renault management . . ., *op. cit.,* 1926.
54. Ehrenburg, *op. cit.,* 1976, p. 23.
55. R. D. Sealey, *From Carriages to Cars,* Oxford: Ruskin College (unpublished Diploma thesis), 1976, pp. 19–21.
56. J. Walcher, *Ford oder Marx. Die praktische Lösung der sozialen Frage,* Berlin: Neuer Dt. Verlag, 1925.
57. P. Berg, *Deutschland und Amerika 1918–1929,* Berlin, 1963, pp. 96–132. R. M. Wik, *Henry Ford and Grass-Roots America,* Ann Arbor: University of Michigan Press, 1972, p. 4.
58. Devinat, *op. cit.,* 1927, pp. 174–211.
59. Castronovo, *op. cit.,* 1971, pp. 327–343.
60. Layton, *op. cit.,* 1974, p. 381.
61. C. S. Maier, *op. cit.,* 1971, p. 55–56.
62. Layton, *op. cit.,* 1974, p. 382.
63. C. Brown, *Sabotage,* Nottingham: Spokesman Books, 1977, pp. 196–201.
64. See, *inter alia,* Marty-Rollan, 'A propos de la rationalisation: les travailleurs devant le progrès', *Le Peuple* **10**, 1 July 1929.
65. P. Saint-Germain, 'La chaîne et le parapluie', *Les Révoltes logiques* **2**, 1976, pp. 87–104.
66. J. B. Jefferys, *The Story of the Engineers 1800–1945,* London: Lawrence and Wishart, 1946, p. 207.

67. D. Montgomery, 'Quels standards? Les ouvriers et la réorganisation de la production aux États-Unis (1900-1920)', *Le Mouvement Social*, January–March 1978.
68. A. Gramsci, *Americanismo e Fordismo*, Milan: Universale Economica, 1950, p. 32.
69. Sealey, *op. cit.*, 1976, p. 21.
70. Castronovo, *op. cit.*, 1971, p. 341.
71. Gramsci, *op. cit.*, 1950.
72. Jefferys, *op. cit.*, 1946, p. 212.
73. National Archives, Paris, F 7 13932 and F 22 185, various reports.
74. A. Clément, 'Fabrication et manutention à la chaîne mobile se développent dans les industries', *L'Humanité-Dimanche*, 17 June 1928.
75. A. Castellani, *Les grèves dans l'industrie automobile de la Seine 1921-1930*, Université Paris VII: unpublished M.A. thesis, 1974, p. 44.
76. Clément, *op. cit.*, 1928.
77. Interview with John Pringle, 9 March 1977.
78. C. S. Maier, *Recasting Bourgeois Europe*, Princeton: University Press, 1975, p. 583.
79. G. Nowell-Smith, *Antonio Gramsci: Selections from the Prison Notebooks*, London: Lawrence and Wishart, 1971, p. 278.
80. W. H. Form, *Blue Collar Stratification*, Princeton: University Press, 1976, p. 137.

IDEOLOGIES OF 'ART' AND 'SCIENCE' IN MEDICINE:

The Transition from Medical Care to the Application of Technique in the British Medical Profession

JUDY SADLER

University of Manchester

This study examines the transformation of English medicine (1) over the last two centuries from a mode of organization based on traditional and craft knowledge to an occupational structure in which science has become the basis of medical practice. This process of 'scientification' (2) of medicine, however, is not one-dimensional and unidirectional A major point of this analysis is that the 'scientification' of medicine (as that of other professions based on the production of knowledge) must be seen as involving, struggles for authority which are both scientific and 'political' in nature (3). There are two aspects to this view which, although being inter-related, may be analytically separated. One is the change in the professional organization (4) of medicine brought about by the development of the science(s) (5) which necessitated a technical division of labour. As a result of this the hierarchical authority structure of the profession was threatened by the breakdown of the traditional boundaries between those considered medically 'superior' and those practitioners of lower standing.

The other aspect is the change in ideologies under the impact of the development of science. I will argue that those in authority (namely those able to control medical practice) were forced to alter their ideologies (6) concerning the nature of medical practice and knowledge under pressure from developing medical knowledge, but that in Britain at least they have succeeded in maintaining an ideology of art in medicine; that is, a conception of medicine as concerning something more than the production and application of science. Thus, though we may talk of the 'scientification' of medical practice, there has not been a concomitant total 'scientification' of medical ideologies (7).

Krohn/Layton/Weingart (eds.), The Dynamics of Science and Technology.
Sociology of the Sciences, Volume II, 1978. 177–215. All Rights Reserved.
Copyright © 1978 by D. Reidel Publishing Company, Dordrecht, Holland.

The adoption of science by the medical profession was mediated by the occupational structure of medical practice and the ideologies of medicine. This is to say that the process of scientification is seen in interdependence with the professional organization and medical ideologies. Although both are subject to change under the impact of scientific development it can also be shown in which ways they have constrained or fuelled the production of scientific knowledge. Finally, it will be argued that the process as a whole, i.e., the scientification of medicine, the challenges to the traditional authorities in the medical profession, the changes in the medical ideologies and, as a result of these, the changes in the distribution of medical care, has to be seen in the context of fundamental changes in the social structure of British society.

1. Preliminary Conceptualization of the Structure of the Medical Profession and of Medical Ideologies

The Production of Medical Care

Jamous and Peloille (8) conceptualize medicine as three productive systems: (i) the production of medical knowledge, (ii) the production of medical care and (iii) the production of medical practitioners. In this view, the same labour force (the medical profession, though not necessarily the same individuals) are involved in the three systems. The systems have a certain degree of autonomy. In (i) doctors work upon 'cognitive objects' to produce new cognitive objects; in (ii) they work upon patients/symptoms/ diseases to produce health; and in (iii) doctors teach recruits to produce new medical practitioners. Those who have 'authority' in the medical profession can be seen as those in possession of the means of production of each of these productive systems. Authority struggles within medicine have centred around this possession and its concomitant control of the three productive systems.

Central (or dominant) to these productive systems is the production of medical care, which is the basis of the economic viability of the other productive systems. That is, it is on the basis of 'selling' medical care that the production and distribution of medical knowledge and medical practitioners is ensured. The production of medical care is the source of income for the profession and thus is the basis of the occupational structure. (9)

The Authority Structure of Medicine

Though the production of medical care is dominant amongst the three productions of medicine, the unity of these productive systems is provided by (iii), the production of medical practitioners. For it is around this activity that the profession has been organized. That is, the profession has traditionally been controlled, and entry restricted, through the professional associations concerned with monopolizing medical care – the source of income. The profession has maintained, as far as possible, control over health practice by claiming to be the only holders of the knowledge (whether craft, art, or science) necessary for the production of medical care. The professional associations are mechanisms for recognizing recruits as possessing this knowledge, whilst restricting numbers in order to monopolize the population requiring health care (in particular the wealthy) and ensuring an adequate income for their members. It is therefore control over the means of production of medical practitioners that has ultimately ensured the profession its control over the distribution of (at least the most important section of) medical care.

Through control over the means of production of medical practitioners, an elite has traditionally controlled the production of medical care. The production of knowledge, on the other hand, was at one time largely irrelevant to the 'relations of production' which ensured the control over access to the treatment of the wealthy by an elite. With the rise in the importance of science and technology throughout society, this position changed, and the production of medical knowledge became integrated into professional organization (10). This transformation was a result of authority struggles within the medical profession in which those seeking control over a section of medical care used science as legitimation for their practice in a time when sources of legitimation in society as a whole had altered. In the modern period, those who control the production of medical practitioners still control the most financially rewarding medical practice, but on the basis that they are 'qualified' to control these productions because they control, and are the producers of, medical knowledge. The production of medical knowledge has thus become legitimation of medical authority and an important concern of the profession.

Ideologies of Medical Practice

Two opposing ideologies concerning the nature of medical knowledge have been central to authority struggles within medicine since the eighteenth century. The dominant ideology at any time has been that postulated by the medical 'elite' who control the production of medical practitioners and medical care, and this ideology has thus also been a *definition of the nature of medicine*. One view, traditionally dominant, postulates that medicine is an *art*, whilst the other postulates that medicine is a *science*. Whilst the precise content of each ideology has changed over time, there are a number of elements that have remained constant in each.

Ideologies of 'art' have been based on the idea of tradition and craft experience. The practice of medicine is seen as involving elements of knowledge which cannot be fully conceptualized or, in Jamous and Peloille's terms (11), are 'indeterminate'. By indeterminancy the latter refer to the 'means' of transmission of mastery of intellectual or material instruments which "escape rules and, at a given historical moment are attributed to virtualities of producers" (12), as opposed to 'technicality' which refers to 'means' which "can be mastered and communicated in the form of rules" (13). Ravetz's definition of craft activity is close to the ideology of 'art' in medicine:

Craftsman's work is done with particular objects, which may be material or intellectual constructs, or a mixture of the two: and the operator must know them in all their particularity. Their properties and behaviour cannot be fully specified in a formal list; in fact, no explicit description can do more than give the first simple elements of their properties. Hence the operator's knowledge of them must be 'intuitive' . . . It cannot be learnt from books, but from experience, derived from a teacher by precept and imagination, and supplemented by the personal experience of the operator himself (14).

There are three elements fundamental to concepts of art in medicine: (i) 'Indeterminancy' or 'intuitive knowledge', (ii) learning from experience and imitation, and (iii) particularity: each patient is specific and cannot be treated like any other. It will be suggested later that the latter is linked with a particular 'mode of production' of medical care, that based on private practice. Concomitant with this ideology of art, we see apprenticeship playing an important part in medical training: this was the traditional method of obtaining craft knowledge.

Ideologies of 'science' stress 'determinate' features. That is, in this conception medicine involves the production and application of precise intellectual structures and techniques which can 'theoretically' be learnt by anyone given the opportunity. The increased importance of this ideology has coincided with the increase of formal medical education (lectures in university setting) and the growth of links between training and the producers of scientific knowledge.

These ideologies have been part of struggles within the medical profession to assert a definition of medicine. The struggles for authority cannot be seen as irrelevant for the sociology of science: they have necessarily had an influence on the production of knowledge. Yet neither should these struggles be seen simply as concerning the nature of knowledge: they are rather inseparably bound with political struggles to control the production and distribution of medical care. As such, they are concerned with struggles for access to financial resources by sections of the profession.

It will be seen that it is by viewing such ideologies as being tied to the production (including distribution) of medical care, that it becomes possible to link them to general societal legitimations and to the role of governments in consolidating medical authority. For linking ideologies of the nature of medicine primarily to the occupational structure permits us to examine the conditions for crisis in the production of medical care as part of the dynamic of general societal changes and to examine the role of science in this process. It will be seen that these ideologies have been studied here not simply for what they reveal about the development of scientific medicine *per se*, but for what they reveal about the position of science within the occupational structure based on the production of medical care.

2. 'Art' and 'Science' in Medicine: Sixteenth to Nineteenth Centuries

Traditional Medical Authority

It has been suggested that the medical profession has been structured around the control over the means of production of medical practitioners by a section, an 'elite', of the profession. This elite was traditionally the physicians who dominated and controlled the surgeons and apothecaries with the backing of the English government.

The composition and size of the labour force for medicine in England has traditionally been controlled by the medical professional associations, dominated by the Royal College of Physicians (RCP). Established in 1518, the charter of the RCP provided for the prosecution and imprisonment of unlicensed medical practitioners (15). The license was granted by the College and the requirements were that candidates should have a good classical liberal education and should then study medicine, which consisted of gaining knowledge of – and ability to translate – the classical medical texts. Candidates then took the College's oral examination. Whilst knowledge was considered necessary for the practice of good physic, *medical* knowledge was not greatly emphasized (16). A liberal education was seen as creating the *qualities of personality* necessary for sensitive medical practice, whilst medical 'know-how' was perceived by physicians as being obtained principally after licensing, through the *experience* of medical practice.

The license was the mechanism for a *social* distinction (the ability to pay for a long education) rather than being primarily concerned with examining competency in medical technique. By the end of the eighteenth century, physicians were more known for their wit and elegance than for their medical knowledge (17). Medicine was perceived as being an *art* involving far more than technique: quality, indeterminancy and experience in the form of cultural sophistication were emphasised. The medical practitioners of technique, the 'mechanical arts', and of trade (the surgeons and apothecaries) were scorned – and carefully controlled – by the physicians (18).

The possession of the means of production of the physicians (which also indirectly involved control over the production of surgeons and apothecaries), enabled the physicians to control the production – and in particular the distribution – of medical care. Through the medical license the total number of physicians was strictly limited in order to ensure an adequate income for licentiates of the College. The licensing system effectively ensured their monopoly over the treatment of the aristocracy and their wealth was gained through patronage: charging fees was condemned as 'mere trade' (19). The ideology of art in medicine was directly related to the source of income of the physicians – the aristocracy – and mirrored that class's condemnation of trade.

Through their links with the aristocracy and the House of Lords, the RCP was able to control the production of medical care by the surgeons and apothecaries through legislation. In effect, the RCP prevented the 'lower orders' of medicine from encroaching on their lucrative section of practice. Through legal statutes, surgeons were permitted only to practise the 'mechanical arts' of surgery, whilst apothecaries were limited to prescribing drugs. However, the physicians failed to prevent apothecaries and surgeons from practising altogether (20). As physicians seldom treated the poor, there was a constant opening for medical practice in this area – whether legal or not. Legal control over the activities of the 'lower orders' was, therefore, all that the physicians managed to obtain. Yet in this period the physicians did have the backing of the government.

The surgeons and apothecaries – also having their livelihood threatened by expansion in the numbers of medical practitioners – formed their own professional associations (latterly named the Royal College of Surgeons and Society of Apothecaries respectively (21).). They too, established licensing as a mechanism for the production of medical practitioners and through this means restricted the number of 'official' medical practitioners.

Each grade of medicine established a restrictive mechanism for the production of medical practitioners *in order to obtain control over a section of medical care.* Physicians were clearly dominant and it was their definition of medical practice, in which medicine was seen as an art, which was hegemonic. The secondary position of surgeons and apothecaries was justified by referring to the mechanical nature of their practice, which was considered 'trade' not 'care'. Though the education of the 'lower orders' was based on apprenticeship and could be said to involve the craft characteristics of 'medicine as an art', the fact remains that in the pre-scientific era of medicine, these practitioners were denigrated for their *mechanistic* and *technical* approach (22).

Though medical science was developing, it was still relatively peripheral to professional concerns in the eighteenth century. As Carr-Saunders and Wilson wrote:

> . . . throughout the 18th century the professions were regarded first and foremost as gentle-men's occupations. Though they might not offer large material rewards, they did provide a safe niche in the social hierarchy (23).

Only the practice of physic was considered a profession: surgeons and apothecaries were traders. Because of the hegemony of the physicians, the dominant emphasis in medicine tended to be on gentlemanliness rather than medical knowledge and technique (24). Indeed Carr-Saunders and Wilson argue that the organization of the medical profession in a hierarchy with the physicians at the top, ". . . *militated against the full employment of the medical knowledge of the time*" (25). In addition, they point out, the division of labour between apothe-caries, surgeons and physicians "*was not dictated by the nature of medical technique*" (26). It was, rather, concerned with control over the production and distribution of medical care, over which the physicians were pre-eminent. In other words, if their view is correct, it is more valuable to see the medical grades as manifestations of a social division of labour than as a technical division of labour resulting from the structure of technique and knowledge of the period.

Change in the Distribution of Medical Care and the Development of Science:

During the eighteenth, and especially the nineteenth, centuries vast changes in the social structure of British society took place. This inevitably affected the rigid hierarchical structure of the medical profession with the physicians at the top, the surgeons in the middle, and the apothecaries at the bottom. In particular, the changes in the social structure altered the distribution of medical care (as a result of transformations in the distribution of wealth) and thus the source of income of the medical profession (27).

The rise in the population from the eighteenth century onwards increased the importance of the surgeons and apothecaries (28). The RCP had always restricted numbers severely in order to ensure an adequate income for its licentiates in treating the aristocracy. With the increase in the population, the physicians did not expand their numbers proportionately, preferring to remain exclusively the doctors of the aristocracy. However, new sources of wealth were available with the rise and growth of the industrial bourgeoisie and, by 1860, surgeon-apothecaries (doctors with licenses from both professional associations) had become their accepted doctors (29). The Royal College of Surgeons (RCS) and the Society of Apothecaries increased in status and a number (though by no means all) grew in wealth. Because changes in the distribution of wealth enabled many surgeons and apothecaries to establish relatively successful practices, their increased importance was inevitable. Yet an attack on the legal control imposed on them by the RCP remained essential for the lower strata of medicine if they were to maintain this position and were to be able to control their own affairs. In particular, they needed to legitimize their position as doctors *competent* to treat the new wealthy. Science was to be central in the struggle against the authority of the physicians.

Standards of legitimacy of what constituted knowledge within society as a whole changed with developments of the industrial revolution. Surgeons had begun to improve their techniques during the eighteenth century in order to obtain the new wealth available to them with the rise in the population. Whilst they had previously been denigrated as practising a 'mere' trade based on mechanical arts, 'trade' gradually took a more important place in society.

This improvement in skills must be seen as an attempt to gain more control over illness. In this sense science must be seen as a genuine development in the 'forces of production' enabling greater success in the production of medical care. The continued development of technique by individual doctors was reinforced by the possibility of establishing successful private practices. But this is by no means a complete picture as it fails to explain the relatively static nature of medical knowledge earlier.

The general development of science since the Renaissance, and particularly since the latter half of the eighteenth century in Britain, is beyond the scope of this study. As far as English medicine is concerned, *the new competition in medical practice was a large, if not the major, factor in the development of scientific medicine.* This was competition between and within the medical ranks for access to patients within a changing social structure in which the distribution of wealth was transformed.

Illustrative of this process was the challenge of the Scottish physicians. The latter were trained in an atmosphere of flourishing scientific development and, though scorned by English physicians because of the laxity of examinations in Scotland, most of them were well-versed in the new scientific medicine of Scotland and Europe.

By the latter half of the eighteenth century, the number of physicians in Scotland was out of proportion with the size of the population and many moved into England to escape the severe competition of medical practice. They entered general practice treating the middle classes. Immediately the livelihood of the surgeon-apothecaries was threatened, for the surgeon-apothecaries had become the general practitioners of the middle class with the failure of the physicians to break their tight links with the aristocracy early enough. As Parry and Parry argue (30), this challenge to the surgeon-apothecaries led the latter to improve their own scientific standards. The apothecaries, in particular, consciously took up science in order to improve their standing in society and to ensure the consolidation of their 'profession', This culminated in the Apothecaries Act of 1815 which, though motivated by the desire to restrict the medical practice of 'quacks',

. . . gave statutory force to the notion of formal professional education under the control of a professional body and tested by strict examination. Licensing was not made a function of the State, but it was backed with the State's authority (31).

The Act prohibited practitioners who had no license from the Society of Apothecaries from calling themselves apothecaries (32). The Society instigated the first modern medical examination (written, not oral) and required certificates of attendance at lectures on anatomy, physiology, medicine, materia medica and chemistry before a licence could be given. The candidates also had to serve a five-year apprenticeship, six months of which was to be spent in a hospital (33). The surgeons, threatened by the possibility that apothecaries might now be considered better qualified than themselves followed suit with their own improved licence requirements.

This Act has significance for a number of reasons. First, in their licence requirements, the apothecaries were asserting the central position of science in medicine. Second, in doing so, apprenticeship and craft experience are no longer perceived as the only means of gaining medical knowledge. Formal medical education in the form of lectures on science takes an important place. Indeed this Act, followed by the action of the surgeons, was to cause both an extension of scientific medical knowledge and the development of private medical schools which began the extension to more formal medical education (34). A number of these schools were to become the basis of the new red brick universities. By 1827 the first modern medical school, University College, London, had been established. Determinancy, science and formal education were encroaching on indeterminancy, art, craft and apprenticeship.

The RCP now had an outdated mode of qualification which it gradually altered under the precedents of the 'lower orders'. Yet, though many have seen the Apothecaries Act as a defeat of the authority of the physicians over the apothecaries, this was probably not the case. The Act was essentially a compromise towards the demands of the physicians, in which several clauses of the original bill had to be dropped and several clauses drafted by the RCP were added, before the proposals finally became law. It was the physicians, not the apothecaries, who still had the ear of the government, and it remained the physicians who controlled the production of medical practitioners and the production and distribution of medical care. Thus, though science was part of efforts by apothecaries to legitimize their medical practice, they were only able to limit the activities of 'quacks'. They neither defeated the legal restrictions imposed by the physicians, nor defeated their definition of medical practice – that medicine was an art.

Science and the Hospitals

It is necessary here to discuss changes in medical knowledge and the institutional conditions for the production of scientific knowledge.

In the traditional private medical care system, control over the means of production meant possession of the means of production (and distribution) of medical care. For a technical division of labour was undeveloped and the means of production of medical care resided with the individual doctor – it involved the doctor's knowledge plus the provision of a small amount of tools, instruments and drugs which could easily be obtained by the individual doctor. The development of medical knowledge was to create both a technical division of labour and the need for the kind of technology in medicine which was beyond the scope of provision of the doctor. The institutional setting for the production of medical care was later to be forced to change under this pressure from being primarily in the private house to the hospital. This will be discussed in the next section. What interests us here is that the hospital was to become (i) the centre of the production of medical care (though not the exclusive place), (ii) the centre of production of medical practitioners (in conjunction with formal teaching centres), and (iii) the production of medical knowledge. Future struggles within the profession were to involve attempts to gain control over the hospitals which became the institutional basis for the control of medicine (35). Essentially the importance of the hospital was due to the move away from conceptions of illness which were specific to the individual (36).

Hospitals became important centres for the medical profession with changes in conceptions of illness. Rosen (37) argues that only with a break from humoral-type pathologies to the belief in disease entities could the pooling of a mass of sick people gain the significance that it did. The ancient conceptions of illness, of which humoral theory and solidism were dominant, were based on ideas of body states in which illnesses were seen in terms of imbalances of the body. Such a view was individualistic and patient specific. The function of the doctor was to control these imbalances.

During the eighteenth and nineteenth centuries, disease entities became widely recognized and the focus of medical conceptions (38) changed. Symptoms were correlated across patients and clusters of symptoms tabulated. Though it was only with the breakdown of humoral type pathologies that hospital populations could gain significance for the production of medical knowledge, Rosen also suggests that the development of disease-entity pathology had awaited the concentration of urban populations in which similar illness conditions could be recognized (39). This culminated in hospital populations of considerable size during the industrial revolution. With the development of science, the universality of illnesses became increasingly recognized and the 'individual', the 'human subject', became increasingly lost within medical knowledge. Although the human subject was to continue to play an important part in the ideology of medical practice later, science moved beyond the patient as person to conceptual structures of increasing remoteness (e.g., biochemistry) from the idea of the human subject.

During the nineteenth century, the hospital became the centre of the production of scientific medical knowledge because of the availablity of large numbers of patients for study. It also became a locus of medical training and the basis of legitimacy for gaining lucrative private practice:

> . . . appointments to hospitals were the best advertisement any physician might have for promoting his private practice, and there were direct financial rewards in the apprenticeship fees to be gained from bedside teaching (40).

The voluntary hospitals for the poor had always been the domain of the physicians and the 'best' surgeons, who held honorary (unpaid) posts (41). Physicians were beginning, through the voluntary hospitals which provided 'specimens' to be studied, to control the production of medical knowledge. Surgeons also had this prerogative, though strictly limited to their field of 'mechanical' practice. Thus those who traditionally controlled the voluntary hospitals (the physicians and most eminent surgeons), and those who had been most reluctant to emphasize science in medicine, became producers of scientific medicine. Yet though they might be producers of medical knowledge, they still tended to emphasize 'art' in medicine, and in particular, the qualities of the human subject which they viewed as irreducible to science. Their art was 'care', not the application of technique. Nevertheless, the wealthy increasingly expected treatment from the producers of scientific knowledge and as time passed they turned to those in control of the hospitals for treatment.

Medical Reform

The control of the hospital was to play an important, though indirect, part in the battle by the medical reformers of the nineteenth century to break the authority of the hierarchically arranged professional associations. By the mid-nineteenth century, the increased importance of the surgeon-apothecary in the social structure made it possible for them to mount a campaign against the ranking system of medicine. The reform movement, centred primarily in the provinces where surgeon-apothecaries were the most prevalent (42), wanted the establishment of a central licensing body covering the whole of the U.K. with a single portal of entry into the medical profession. Though the movement was really concerned with destroying the power of the medical corporations, particularly the dominance of the physicians, they argued that it was the medical corporations' control over medical practice that was preventing the improvement of medical skills and knowledge (43). The reformers took up science against the art of the Royal Colleges which, they claimed, were lax in the assertion of standards for licensing.

The RCP and the RCS (especially the former) were opposed to any change. This was principally because the present system of distinction between medical ranks prevented apothecaries gaining honorary posts in the charitable voluntary hospitals. The surgeon-apothecaries, the most prevalent doctors, sought entry to these institutions. Specifically they wanted 'merit', which they equated with technical competency, to be the determinant of status within the profession, rather than wealth or birth (44). Once again subordinate sections of the medical profession attempted to assert a definition of medical activity which meant science. Only when surgeon-apothecaries gained 'equality' in conceptions of competence to practice could they hope to gain access to the voluntary hospitals.

When the government finally intervened in the reform struggle that had been going on for many years, it was with a compromise. Though legitimizing the standards and abilities of the 'lower ranks' of medicine, whose role in improving medicine and particularly in reducing deaths at child-birth was widely recognized (45), the legislation compromised the demands of the medical reform movement to pressures from the physicians. The Medical Act of 1815 provided for the establishment of a medical register under the care of the General Medical Council (with representatives from the professional associations, universities and the Crown), which was responsible for the maintenance of academic standards. Licensing however was left in the hands of the various professional bodies: the professional associations and the new universities which had recently gained the right to give medical licences (46).

Though the reformers had claimed that their movement was concerned with knowledge and the need for medical practitioners to have an adequate training, in fact negotiations for the Medical Act ". . . had nothing to do with the promotion of science. Little indeed had been heard about science . . ." (47). It was a battle between the 'lower' and 'higher' ranks of medicine in which science had legitimized the 'lower orders'. Through partially changing the mechanism of production of medical practitioners (from emphasis on the license to emphasis on the register), the RCP had finally lost its hegemony over medicine. For all doctors were equal on the register. Yet the reform movement, which had wanted a single portal of entry into the profession (thus breaking the power of any of the professional associations over sections of medicine) had also lost.

The Medical Acts represented a transformation in the balance of authority within medicine. It has been suggested that this change had been related to changes in the structure of British society (increased population, the rise of the bourgeoisie, etc.), which had impinged on the competitive process of medical practice. Science had played an important part in overthrowing the aristocratic hegemony over medicine, an hegemony which had probably inhibited the use of science. It was authority struggles (struggles to control the means of production and distribution of medical care) that had largely fuelled the wider utilization of scientific developments within the profession.

Traditional knowledge in medicine had been largely of two types: (i) that based on classical medical texts predominantly and (ii) traditional craft-type knowledge gained from experience and passed on from generation to generation. By the time of the Medical Acts, such a conception of medical knowledge had been destroyed. Medicine had become science-based in the sense of utilizing knowledge that was constantly changing and developing as the conceptual structure of knowledge was worked upon. Though the Medical Acts were a result of a struggle over access to patients (control over the production of medical care), not over science, the establishment of a licensing system regulated by the state (48) was *the legislative recognition that scientific knowledge had become essential for the production of medical care.* The General Medical Council was to judge the standards of medical examinations, now written and almost exclusively scientific.

In line with the increased importance of science, medical education was becoming formalized, though apprenticeship still remained important. Medical schools, more and more linked with hospitals, and the new universities, were providing lectures on scientific knowledge. They were also centres of the production of medical practitioners, as students spent part of their training 'walking the wards' with a physician or surgeon in order to learn clinical medicine (49).

The Reconsolidation of Medical Authority

Despite these changes, by leaving the professional associations intact, the Medical Acts left open the possibility of the re-establishment of medical authority. Control over the hospitals was to be strengthened by a new system of qualification in which this control was to be legitimated by scientific knowledge – and increasingly on the authority of being scientific producers – *though still strongly influenced by the concept of medicine as an art.*

After the Medical Acts, the RCP and the RCS established a joint medical licensing system based on a common examining board for examinations in physic and surgery. This involved a deliberate exclusion of the Apothecaries' Society which was still denigrated by the Royal Colleges. As a result, the apothecaries examination – which could not include a paper on surgery and was not a complete medical qualification – became redundant and the Society of Apothecaries declined in importance.

The Royal Colleges restructured their internal organization by instigating 'postgraduate' ranks. The RCP created the rank of Membership of the Royal College of Physicians (MRCP), obtained by passing an examination. Only members and not the licentiates could be selected for the Fellowship (FRCP) and, as the Fellows governed the College, the higher examination was a first step to influence on College affairs. The RCS created the grade of fellowship (FRCS), entry by examination, ruling that only fellows could be elected to the governing body of the College.

The pass rates were, and remained, small for the MRCP and FRCS (e.g., in 1933, 77 candidates attempted the MRCP. Only eight passed.) (50). As Stevens suggests:

The concept of rank, the lower and higher category of doctor, was thus retained, but ironically it was now within the Royal Colleges (51).

The MRCP and the FRCS, requisites for control of the colleges, now also became necessary for voluntary hospital appointments. The MRCP and FRCS holders thus replaced the old position of the physicians in the medical authority structure: licentiates took the apothecaries' place. Concomitantly, the MRCP's and the FRCS's became hospital consultants, whilst licentiates became general practitioners.

Through the MRCP and the FRCS, an 'elite' was able to control the labour force of medicine more effectively than ever before. The higher echelons now held control over (i) the production of medical practitioners through licensing, giving higher qualifications, and through hospital teaching, (ii) the production of medical care: they monopolised the wealthiest patients who sought care by honorary hospital doctors and (iii) the production of medical knowledge, which took place mainly in the hospitals. Even the scientific composition of the medical profession could be, and was, constrained through the higher examination structure, there being considerable financial disadvantage for those who produced knowledge or practised in a field not recognized by the dominant doctors (in particular, those areas not associated with the most lucrative private practice). This will be examined in the next section. However, despite the reconsolidation of medical authority, one feature of medicine was now distinctly changed from the beginning of the eighteenth century: the body of knowledge was now scientific and open to development and change.

3. Art and Science: Scientific Generalism Versus Specialism

By the mid-nineteenth century, physicians, surgeons and apothecaries were concerned with the development of medical technique based on scientific knowledge. I have suggested that this was intertwined with a search for legitimation by the surgeons and apothecaries in which they sought to assert their ability to control sickness by showing scientific 'credentials'. In this section I will examine how the development of scientific medicine created a crisis in the 'mode of production' of medical care, that based on private practice.

Medical Specialization

By the nineteenth century, illnesses were no longer patient specific. Once medical knowledge had passed from the human subject to disease entities, the path was opened up to the development of medical specialization in the production of medical care. At first, this took the form of focalization on parts of the body (e.g., the heart: cardiology) or on specific social groups (e.g., children: paediatrics). Then more conceptually differentiated sciences such as medical chemistry, biochemistry and radiology developed (52).

It has been pointed out that an 'elite' of the Royal Colleges had, since the latter half of the nineteenth century, been able to control the production and distribution of medical care through their possession of the means of production of medical practitioners at the 'postgraduate' level. Scientific specialism now posed a threat to the higher ranks of the Royal Colleges by attacking the basis of the very institutions through which control over the voluntary hospitals and the medical profession was asserted.

Specialism destroyed the boundaries between the Royal Colleges which, as was previously suggested, was based more on control over medical practice than on any necessary technical division of labour (53). Many of the new specialities covered both physic and surgery (e.g., paediatrics, opthalmology). If such knowledge fields were extended and developed, the examination structure of the FRCS and FRCP (confined to the skills necessary for surgery and physic respectively) would be 'out of line' with the body of knowledge and practice. Specialization therefore threatened not only the mechanism of production and distribution of medical practitioners (the higher qualifications), but also the Royal Colleges and their governing bodies. In addition those eminent doctors already established in traditional general private practice feared that they might lose custom to specialists.

Specialism was strongly resisted by the governing bodies of the Royal

Colleges who attempted to control the knowledge and concomitant practice of the medical profession through their higher qualifications. Their actions were justified by a new ideology of 'art' in medicine which, though accepting science, postulated *scientific generalism*. This ideology was dominant well into the twentieth century and was utilized to ensure that candidates for the MRCP and FRCS were forced to gain knowledge and experience in *all* fields of medical science and practice deemed relevant to the governing bodies of the Colleges if they hoped to pass the examinations.

Control Over the Means of Production of Medical Practitioners and the Assertion of Art in Medicine

Through the MRCP and the FRCS examinations, those who controlled the Royal Colleges (and who also controlled the voluntary hospitals and the upper reaches on private practice), possessed the institutional mechanism through which to posit the dominant definition of medical practice. This 'definition' of medical activity involved a definition of legitimate knowledge and also, by implication, an assertion of the type of knowledge that it was most valuable to produce.

Candidates for the MRCP and FRCS were required to answer questions on all fields of medical practice: art and generalism were the characteristics of the definition of medical knowledge. In reality, the ideology was not truly general. *Generalism* began to mean, more and more as knowledge increased, a wide knowledge of a *limited* section of medical practice. Those areas of medicine not associated with private practice (e.g., radiology, pathology, anaesthology, venereology, public health and sanitation, TB, chest diseases, insanity) were rarely examined so that those working in these fields found it extremely difficult to pass the higher medical examinations. They remained therefore in a subordinate position within the medical profession.

The 'postgraduate' qualifications of the Royal Colleges were mechanisms to maintain control of the voluntary hospitals and the lucrative private practice that was associated with the hospitals' honorary posts. *Thus the examinations with their limited content postulated the kind of knowledge that it was relevant to possess and produce and tended to inhibit the development of knowledge in certain fields.* The necessity for a doctor to obtain a long general training meant that specialization was impossible until late in life: neither therefore could these doctors play a very significant part in the production of knowledge until the completion of training. For the production of knowledge was becoming increasingly focused on specialist fields. In addition, there was a clear financial disincentive from pursuing specialist fields in which private practices could not be established and in which there was a dependence on meagre salaries from the hospitals or public authorities.

Struggles for Control Over the Production of Medical Care

Three main groups have been the centre of struggles for authority since the mid-nineteenth century (54). It is impossible to examine these struggles in depth here, but they will be briefly outlined in order to show how (a) each group was associated with a particular definition of medicine and (b) each struggle has centred on access to finances and patients and has therefore been concerned with the production and distribution of medical care, rather than concerning the nature of medical knowledge. These struggles centred around the voluntary hospitals and the inability of certain groups to be legitimated because they were unable to pass the MRCP/FRCS examinations and/or gain access to honorary hospital posts.

    The development of scientific knowledge was part of the continuing attempt to control sickness. By looking at these three groups of medical practitioner it is possible to see that the 'medical elite', though *restraining* the development of knowledge in certain fields, did not – and possibly could not – *prevent* the development of science in these fields. Indeed, in some cases, although an area of medical knowledge was looked down upon by medical authority, it was nevertheless perceived as *necessary* for the successful medical practice of 'elite' practitioners (e.g., pathology, radiology). In other cases, the medical elite could not prevent the extension of areas of medical practice because of the demand by the wealthy for treatment by practitioners in these fields. In both cases, it was increased control over illness with the increase in knowledge that was the linking factor: specifically the ability to treat, relatively successfully, the illnesses of the middle classes. *The production of knowledge has been mediated by the competitive process for access to patients and has, there-fore, also been directly associated with the distribution of wealth, and concomitant distribution of types of illness, within society.*

With the rise of specialization, those areas of science associated with illnesses most commonly found in the wealthy (or not associated with the poor) found the easiest route to acceptance by the medical 'elite'. Those who developed knowledge in these fields and based their practice on them, though bitterly opposed at first, were the only group of the three groups under discussion to become integrated into the elite prior to the National Health Service (NHS). First, however, they were to wage a struggle for recognition which was to centre around the control of the voluntary hospitals.

In the last half of the nineteenth century, a number of doctors – unable to obtain posts in the voluntary hospitals – established 'special hospitals' with the aim of increasing knowledge in specialist fields and providing treatment for some of the poor excluded from the general voluntary hospitals (55). These specialist hospitals were both (a) an attempt to over-come the possession of the lucrative private practice, medical training, and the means of production of medical knowledge in the hands of the physicians and surgeons who controlled the voluntary hospitals, and (b) an attempt to develop new fields of knowledge and therefore to assert a new conception of medical knowledge (56).

The ability of these specialists to establish themselves as competent medical practitioners and, eventually, to be integrated into the medical elite, was dependent on the interest shown by the wealthy. For the main weight of medical opinion was for many years opposed to the 'unethical' specialist practitioners. The 'elite' in medicine vehemently opposed specialization as being mere science and mechanical intervention in illness.

Yet it was the association of specialists with science that appealed to the middle class (57). They sought out specialist medical care and were enthusiastic to sponsor the establishment of specialist voluntary hospitals. Specialists gained from the widespread demand of the middle class for specialist treatment and also from the enthusiasm of the new wealthy to increase their prestige by having their names associated with sponsoring charitable ventures such as hospitals. Specialist practitioners in turn cultivated an image of themselves as experts in special fields. It was their science, in their view, which made them superior to the general practitioners.

Thus the successful development of specialist science and medical practice was largely dependent of *the success of the production and distribution of specialist medical care for the wealthy*. Despite the regular condemnation of the 'public' popularity of specialist medical care (58) by professional spokesmen, specialization could not be prevented by the medical establishment. Finally many of the old 'elite' became specialists and accepted the large financial gains that could be made as specialists. By this time specialization was being recognized as a *necessity* for one very practical reason: knowledge had expanded so greatly that it was becoming impossible for individual medical practitioners to learn all the knowledge available (59).

It is significant however that the eventual acceptance of specialization was predicated on the rejection of medicine as a science (previously associated with specialization) and the assertion of the need to gain a wide general knowledge of medicine before specializing. Thus the MRCP and FRCS remained general examinations – though testing mainly specialties linked to private practice – and the indeterminate qualities typical of art were again emphasized by specialists and generalists alike. As one specialist wrote:

The important discoveries . . . in this field in the last 20 years have come from those who devoted the chief portion of their time to special work. The moral of this argument, however, appears to be that because any particular branch of medicine or surgery requires aquaintance with particular methods and conditions and dexterity with special apparatus which are not found with the average surgeon, that does not justify any medical man who has spent a certain time in that special study considering himself a specialist. The whole is greater than the parts, and a devotion to special studies and methods should only be superimposed on a good general practical training, and not undertaken from the outset of the surgeon's career. From a good general surgeon, by concentration of attention and effort in a limited field, a competent specialist may evolve (60).

By postulating such an ideology it remained possible, through the higher examinations, to maintain control of medical practice. The boundary problem of the Royal Colleges was neatly solved by dividing up specialties between them.

The link between the acceptance of areas of medical knowledge and practice with the ability of these practitioners to control the distribution of medical care, is illustrated by the comparison with the other two groups of interest here. These groups are (i) Specialists associated with hospital technology and (ii) Public hospital and public health doctors. The common denominator between there two sections of the medical profession was their inability to establish private practices.

Technical specialties such a radiology, anaesthology, and pathology, required technical facilities which were usually beyond the ability of the individual doctor to purchase. These doctors were forced to rely on hospital facilities and, as private practice was home, not hospital, based they were unable to establish private practice. Though they often fought for a fees-for-service system (i.e., private practice relations), they were relegated to the position of 'back-room boys' or 'technicians' dependent on very meagre salaries. These specialists were largely unable to gain the necessary clinical experience to pass the MRCP and FRCS examinations. As such, they were unable to gain 'honorary' hospital posts, which meant they had no say in the running of the hospitals, though they were employed by the hospitals on a salaried basis. They suffered low pay and very poor facilities. Rejected for their technicality and condemned as mere technicians, these specialists were nevertheless symptomatic of a technical division of labour arising out of the development of science. For, despite their low position within the medical profession, their work was soon seen as a *vital support* for the clinical work of those involved in private practice. Yet despite their increasing importance to medical practice, these specialties were unable to increase their status and their history has been marked by an almost continuous struggle for equality with the 'honorary' hospital doctors (61).

It is impossible here to examine the public sector of medicine before the instigation of the NHS. Suffice it to say that there were hospitals under the Poor Law and local authorities. These doctors cared for those not catered for by the voluntary hospital system, for the latter tended to deal only with poor patients who could relatively easily and quickly be treated. The public sector doctors were of low standing in the profession, did not generally have private practices and were salaried employees. Because they concentrated on areas of medicine not usually the focus of the voluntary hospital system (62), they too found it extremely difficult to pass the MRCP or FRCS examinations which tested mainly voluntary hospital work. Their low status was consolidated by their lack of qualifications. Despite this, in the latter half of the nineteenth century onwards, these doctors consciously improved their scientific knowledge of their fields of practice in an attempt to improve their status.

Both the public health doctors and the technical specialists postulated scientific medicine by basing their claim for recognition on their scientific standards. Yet they were constantly thwarted in their struggle for status by (i) their inability to establish successful private practices and their concomitant relative poverty, (ii) their inability to pass the higher medical examinations and (iii) in the case of the public health doctors, the punitive attitude of the Poor Law authorities towards the poor sick made good medicine difficult to practice. These latter doctors, who already worked in the public sector, were amongst the strongest voices pushing for complete state intervention in medicine which they saw not only as a means of improving their own status and financial position, but also as providing the medical skills and facilities for the poor which the latter had always lacked. It should be added here that in the case of the technical specialists and public health doctors, lack of space, equipment and time for research as a result of their poor (often atrocious) employment conditions must have hindered the development of knowledge in these fields.

Ideologies of Medical Practice and Control Over the Production of Medical Care

Technical specialists and public health doctors asserted scientific medicine. They were also often strong proponents of state intervention in medicine. I would suggest that this is not accidental. From the time of the dominance of the physicians to the present day, *the dominant view of the nature of medical practice and knowledge has been related to the occupational structure (the source of income) rather than to the state of medical knowledge.* In the aristocratic period, the assertion of medicine as an art was a mirror of the values of the aristocracy whom the dominant medical practitioners served. We have also noted that initial acceptance of some specialization later was due to the establishment of lucrative specialist private practices, resulting from the demands of the middle class for specialist treatment. Those unable to establish private practices remained in a low position due to the restructuring of the profession through the MRCP and FRCS examinations. I will briefly examine the modern ideology of art which legitimated the MRCP and FRCS structure (63). We will see how this was associated with the 'mode of production' of medical care.

In the latter half of the nineteenth century and the first decade of the twentieth century, the assertion of medicine as an art had been central to the abhorrence of specialization. Specialization was associated with technicality and was thought to deny all the qualities seen as essential to the nature of medicine. Though medical practice was now based predominantly on science – experience, craft activity and indeterminacy in medicine were all stressed in the dominant conception of medical practice. Medicine was

. . . the art of healing. Medicine is so defined by Aristotle and the art of curing, to which we might add the art of preventing, disease has all the characters of an art. It depends on experience and skill, it deals with individual cases, and the perfection it aims at is practical, not speculative – the knowledge how to do and not knowledge how things happen. Nevertheless, as practical navigation is founded on astronomy, meteorology on botany, geology and vegetable physiology, so the art of medicine depends on the science of pathology, the practice of physic on the principle of physic (64).

The irreducibility of the human subject to scientific knowledge was asserted. As for example

In the patient there is that mysterious pye which can never be evaluated – namely the quality or condition known as life – and this is apt to upset all results which should happen according to the teachings of pathology (65).

According to Fortesque Fox, medicine is an art

. . . resting not alone on accepted principles of time, but upon a much wider and more uncertain basis (66).

Though by this time, 1910, it is less clear whether science was simply insufficiently developed. He continues:

Art is called upon to occupy a field coextensive with life, far too great for science to define yet. It is a proud justification of our part that again and again after many centuries science in a later time has established and confirmed its methods upon a rational and a permanent basis. In medicine, practice has always been in advance of theory (67).

Yet the indeterminacy of art in medicine remained a powerful force. For Sir Dyce Duckworth in 1913, for example, the physiologist

. . . is not a bedside artist and has little knowledge of the inward human elements in the ailing human individual. He is often devoid of all clinical instincts. His work is impersonal and confined to the soulless materials. The physician ministers to the mind and soul as well as to the tissues and secretions (68).

He adds:

I have a strong opinion that the best practice in medicine will never be completely covered by what we recognise as science (69).

The early rejection of specialization was closely linked to these conceptions, particularly in the condemnation of the neglect of the human subject in specialist science. *General scientific knowledge* was stressed as essential to good medicine:

With the growing mass and complexity of human needs, the demand for specialization in studies runs far ahead of scientific necessity. It becomes more desirable that in educated circles the theoretic interest – the observant outlook over knowledge as a whole – should be maintained. The specialist who loses note of interest in human interests outside his special field loses something of his humanity. For the physician such a result would be deplorable. The traditions of the profession are, happily, against it (70).

In the same year 1906, the *Lancet* attacks specialization with a statement reminiscent of a century earlier:

It is incumbent on all those who practice medicine to emphasise the fact that our calling is not mere trade (71).

Thus specialization before its acceptance in medicine was associated with 'mere' technicality, a rejection of the human subject, and a failure to recognize that medicine cannot be reduced purely to scientific knowledge. Science in medicine was seen as an essential tool, but a tool that had no value without the qualities of talent of the doctor. Once again, the medical profession was concerned with producing the right type of person to practice medicine, rather than producing medical practitioners with scientific knowledge (72). Even when specialization became predominant, it was only acceptable in a doctor who had already gained a wide general experience in medicine (guaranteed by passing the higher medical examinations) which ensured the development of the necessary qualities in the doctor. The dominant view stressed art in medical practice (73).

This ideology is inextricably linked with the 'mode of production' of medical care based on private practice. Central to private practice is the idea of the 'patient-doctor' relationship. This relationship stresses the indeterminate 'something' which is medical care and which concerns the development of empathy between the doctor and the patient. It is both a rejection of the loss of the human subject within medical knowledge and a direct attack ˙on science that neglects the individual. The doctor, the popular medical press of the early twentieth century states over and over again, treats the patient not the disease. For example, Rosen as late as 1944 asserts,

The physician always deals with sick people, never with diseases (74).

And he continues to state that the effect of science and its concomitant specialisation in medicine is

. . . to weaken the patient-physician relationship. It replaces a single relationship between a patient and a physician by a multiplicity of contacts between several doctors (75).

Stressing indeterminancy, Rosen also says:

Every activity that is constantly repeated gradually becomes mechanical, and the physician who is a complete specialist often tends to this mechanisation. Such a physician frequently loses interest in everything outside his special field. Even the most conscientious physician will not remain free from this mechanisation if he does not consider the entire personality of the patient. *The consequence of this mechanisation is the depersonalisation of the relation between the physician and the patient (76).*

Here we return to the old bogey: medicine as mere technique. More important, however, is the fact that *medical science, and its concomitant specialisation, posed practical professional problems that arose directly from the private medical care system in which the 'patient-doctor relationship' was central and in which teamwork (or technical division of labour) was largely impossible.* In this system the individual doctor produced a product, 'health care', which was sold to individual patients: it was a one to one market relationship between doctor and patient. Specialization was finally accepted because of the ability to 'sell' specialist medical care successfully. But specialization produced one major problem arising from the individual patient-doctor relationship: How was a patient to select the correct specialist doctor without being able first to diagnose his or her own illness? Teamwork was necessitated by medical specialization, yet precluded by the 'mode of production' of medical care. The answer on the level of ideology was to deny the advantages (or at least to pose the limitations) of specialist knowledge by asserting that medical care was something above and beyond the application of technique. The 'practical' solution was not found until the development of the general practitioner/consultant referral system with the provision of a secure income for the general practitioner by the state (77). This 'saved' the patient-doctor relationship: each patient would first see a general practitioner and then, if necessary, be directed to a suitable specialist (both doctors would receive some kind of fee).

Science and Crisis in the Poduction of Medical Care

Yet despite the ability of the medical elite to assert (through the higher examinations) a definition of medical practice and medical knowledge that upheld the requirements of private practice, the production of science was nevertheless threatening the British private medical care system. For the development of science had not only created the necessity of a technical division of labour, but it was also requiring more capitalization for the production of medical care. For concomitant with the crisis in the patient-doctor relationship was the crisis arising out of the inability of the private medical care system to provide the means of production for the developing scientific medicine.

Whilst the development of medical knowledge had centred around the hospital and the treatment of the poor, private medical care was traditionally centred on the home. The private house became increasingly unsuitable for medical practice because of the facilities now required for medical work (e.g., operation theatres). Though there was a gradual development of paying wards (78) in the voluntary hospitals either (i) the hospitals were unable to provide all the equipment now required for good medicine, particularly laboratories and radiological dapartments, or (ii) where such facilities were provided, the hospitals were in massive debt (79). The financial crisis of the voluntary hospitals became so acute that it was eventually the most eminent doctors, those who controlled the voluntary hospitals, that were to be the strongest force urging state intervention in medicine.

Since the mid-nineteenth century, the production of medical care, the production of medical practitioners and the production of medical knowledge, had become closely linked in the voluntary hospitals. Control of these productions, therefore, resided in the ability to control entry into the voluntary hospital posts through the production and distribution of medical practitioners (through the College examinations). The financial danger to the voluntary hospitals was another threat to the medical authority structure.

Sections of the 'lower strata' of medicine had long been pushing for state medicine. They had suffered from inadequate salaries and even more from lack of facilities for their practice and for the development of knowledge. By the Second World War, *all* hospital doctors were suffering from the inadequacies of the charitable hospital service and, for the elite clinicians, the need to provide the middle class with good hospital medical care was increasingly urgent. (80). The 'elite' and the 'lower orders' (though not the general practitioners outside the hospitals) were united in wanting state intervention.

Though the elite had succeeded in asserting a definition of medical practice *beyond* the application of science, but based on science (scientific generalism and indeterminacy), the continued development of the various sciences was overturning the private medical care system and the institutional control of the elite. They were temporarily allied with those who represented scientific medicine in wanting state finance.

When state intervention came, it was not wholly as the 'elite' desired. For whilst government funds for medicine had been envisaged, nationalization of the voluntary hospitals had not. Yet by this time the voluntary hospitals had reached such a state of financial insolvency that, despite nationalization, it was 'elite' hospital doctors represented by the Royal Colleges who were instrumental in establishing the National Health

Service (NHS). This was despite opposition from the 'lower order' general practitioners who wanted the medical profession to refuse cooperation with the new service (81). The Royal Colleges negotiated with the government despite this opposition and gained what they wanted at the expense of the general practitioners: they obtained a state hospital medical service in which private practice could continue. Hospitals in the NHS were provided with private beds. Consultants would bring their private patients into the hospitals and charge private fees for treatment, whilst utilizing the means of production of medical care (laboratories, operating theatres, etc.) and labour force (nurses, porters, etc.) provided by the state. Thus the crisis of the private medical care system was overcome by state intervention, with the maintenance of the most lucrative private practice. The structure of the NHS, particularly with the 'firm' organisation in the hospitals (82), was to indicate the continued dominance of private practice in the professional organization.

It is impossible to examine the NHS in detail here, but it is valuable to note that, though the state service secured income and facilities for the 'lower medical orders', control over the production of medical care, of medical practitioners and medical knowledge, became even more closely held in the hands of an 'elite'. The Royal Colleges (controlled by fellows who were hospital consultants) held positions on all the major NHS committees. They were on hospital appointments committees, advisory committees, The Distinction Awards Committee, and on salary negotiating bodies (83).

The creation of the NHS, which encapsulated and reinforced the traditional medical elite, was not to end authority struggles between various sections of the profession represented again by debates between science and art (now specialism versus scientific generalism) in medicine. The MRCP and FRCS, which continued to test a limited section of medical knowledge under the guise of generalism, was to be a constant source of friction – especially as control over the production of hospital consultants was closely linked with decision-making ability in the NHS. Eventually radiologists, pathologists, and psychiatrists rebelled by forming their own Colleges with their own higher examinations, thus controlling the production of the labour force for their own respective specialties. This finally enabled them to gain representation within the

NHS structure. It had become possible for them to establish their own mechanisms of professional control because state intervention in medicine, with provision of facilities and new posts, had enabled these specialties to expand in size: the provision of medical care had become increasingly dependent on them. Pathology, for example, became the fourth largest specialty in medicine after the Second World War.

Though the NHS structure had consolidated the medical authority structure, state intervention in medicine gave a massive boost to the expansion of scientific medicine – especially in such fields as biochemistry, immunology, cell biology, and radiation techniques. Science education for the undergraduate medical student increased rapidly. In a number of areas the medical profession, which was often incapable of providing the scientific talent required, found 'pure' scientists (e.g., biochemists, physicists) encroaching on the medical domain. These factors have increased the 'formal', as opposed to apprenticeship, aspects of medical education. In recent years, medical students are obtaining additional opportunities for taking a pure science degree (BSc) as part of their medical training. Apprenticeship now plays its most important role in post-graduate training and even here merit is becoming more closely associated with some of the more formal aspects of the modern educational system. Publishing papers has become an important mechanism for obtaining recognition in the profession, indicating further the importance of the production of medical knowledge in the modern professional structure.

Whether private medical care will disappear from the British state medical service is debateable. If it does, it may open the door to the break-down of the traditional 'patient-doctor relationship' in which one doctor has the responsibility for 'his' or 'her' patients. Teamwork (by this I mean an institutionalized technical division of labour), in which the relationship of a number of specialists to a patient may be more in accord with the structure of modern medical knowledge, might become the norm rather than the exception (84). (We might even see the disappearance of the hospital 'firm' and the discontents it engenders in junior hospital doctors.) Those features of the individual typically stressed by the hospital 'elite' may become more the domain of the general practitioner and community physician as the stress in funding policy in the NHS moves away from hospital care to social and preventative medicine. However, even if this is the case, we might expect 'scientification' (determinancy) in these areas: the difference being that social, psychological, and environmental variables would be utilized in the scientific field. For, as Powles stresses, the 'doctor-patient relationship' has had a strong influence on the type of science developed in medicine and its disappearance might give impetus to new scientific fields:

Given the traditional form of doctor patient interaction, it was inevitable that doctors would strive to get better at intervening in their patient's illnesses. This is the historical foundation of the engineering approach. When doctors drew from the emerging biological science of the 19th century, they chose that strand which had most relevance to their ability to treat their patients: They chose what Crombie refers to as the 'science of the organised individual' and were singularly uninfluenced by the other strand – 'the science of populations'. . . . The extent to which human population biology – for example evolutionary theory, historical demography and medical ecology – has failed to influence medical theory is quite remarkable. The resulting inability to deal theoretically (as distinct from statistically) with biological phenomena above a single organism has left medical theory seriously deficient. Medicine has deprived itself of the only basis on which criteria for biological normality in man could rest (85).

4. Conclusion

This paper has suggested that any comprehension of the 'scientification' of medicine must include examination of struggles for authority within the profession. These struggles have concerned the ability to control (i) the production of medical knowledge, (ii) the production of medical care and (iii) the production of medical practitioners. We have seen that possession of the means of production of medical practitioners has been central to the control of the other two productions. Yet it has been (ii) the production of medical care, which is the source of income for the profession, which has provided the rationale behind authority struggles. Examination of this occupational function has enabled us to link medical authority struggles with factors 'external' to medicine. Changes in the social structure (including culture) of the population requiring medical care have played an important role in determining the outcome of these struggles, primarily by introducing changes in the structure of distribution of medical care.

Since the end of the eighteenth century, science has been utilized by sections of the medical profession to legitimize their claim for access into the 'elite' strata of medicine. This does not mean that science was 'merely' an ideology: the fact that these sections believed that they *could* legitimize their medical practice through science suggests that it had also become, to some extent, the basis of their practice. There is little doubt however that the dominant conceptions of the nature of medical practice had more dependence on the 'mode of production' of medical care than on the structure of medical knowledge *per se*. Thus, the development and structure of medical knowledge has been influenced by occupational priorities (i.e., access to sources of income found in the provision of medical care) rather than simply concerning the process of working on problems 'posed' by the conceptual framework of medical knowledge. On the other hand, science has produced a technical division of labour, the structure of which is open to alteration with the development of science, which may question the authority structure of the medical profession organized around the production of medical care.

Notes and References

1. This study concentrates on English medicine because of divergences in history between Irish, Scottish and English strands of development.
2. The concept of 'scientification' in itself poses many problems. However, in the context of medicine I have used it to mean the transformation from relatively static knowledge based on classical medical texts and folk remedies, to a period in which medical knowledge was perceived as open to constant change and in which the production of medical knowledge became institutionalized into the professional structure. This does not mean that I do not recognize that there were developments in knowledge in earlier periods. However, the extent to which these developments became widely utilized in practice was limited. It should also be mentioned here that although I refer to the 'development of science', I am not suggesting that science is a unified structure that develops in a linear fashion, but use it only as a convenient shorthand.
3. P. Bourdieu, 'The Specificity of the Scientific Field and the Scientific Conditions of the Progress of Reason', *Social Science Information* **14**, 1975, pp. 19–47.
4. It will be noticeable that this paper avoids the usual Anglo-American approaches to the 'professions', and could be described as being within a more European mould. (The work of T. J. Johnson and of H. Jamous and B. Peloille can be seen as typical of this mould.) In particular, concepts of 'professionalization', 'institutionalization of the profession', and 'monopolization strategies' are not utilized. The inadequacies of these most common conceptualizations have been lucidly examined by Johnson (T. J. Johnson, 'Professions and Power', *Studies in Sociology*, Macmillan Press 1975) and do require repetition here. Whilst the approach of this paper does not claim to be without its own problems, it does at least avoid some of the pitfalls typified by Berlant's work (J. L. Berlant, *Profession and Monopoly: A Study of Medicine in the United States and Great Britain*, Berkeley, Los Angeles: University of California Press, 1975). By viewing

the history of medical professions as a process of institutionalization in which Weberian theory of monopolization is viewed as the most valuable, Berlant falls into the trap of ignoring many important features in an attempt to fit the 'facts' to the 'theory' of monopolization. (My criticism here is restricted to his work on the British medical profession.) Though Berlant provides some interesting insights into the British medical profession and its reaction to important ideologies (e.g., liberalism), his theoretical approach has necessitated viewing the medical profession as a body with collective interests which has produced 'monopolization strategies'. Consequently, Berlant has seen the development of British state medicine as a result of a particular monopolization strategy of the profession. Because the British state would not make medical practice illegal for non-state-registered doctors, Berlant argues, the medical profession took up a strategy of utilizing state medical services which restricted the participation of practitioners to licensed doctors and therefore enabled them largely to control the medical market. To Berlant, ". . . the use of state employment and financing to induce patients to use 'free' NHS practitioners represents the ultimate logical conclusion of the new monopolization strategy" (Berlant, *op. cit.*, 1975, note 10, p. 169). There are two inadequacies with this view. Firstly, as I shall show later, the medical profession has rarely possessed 'collective interests', but has been rent with divisions in which different sections have been competing for access to the medical market. Any 'monopolization strategies' would therefore have to be seen in relation to authority struggles *within* the profession. Secondly, and perhaps most important, Berlant ignores the importance of the development of science within medicine. As will be shown, it was the inability of the private medical care system to provide the expensive technology required for medicine with the development of the sciences that created a crisis in medicine and caused the most powerful medical practitioners to turn towards the state provision of funds.

5. See note 2.
6. In using the concept 'ideology', I am referring to the ideas held by particular social groups about medical practice and knowledge. I am not making any statements about a "distortion, or reflection, of the real".
7. That is, there has not been a development of the belief that science is the only valid knowledge for medicine throughout the profession.
8. H. Jamous, and B. Peloille, 'Professions as Self-Perpetuating Systems: Changes in the French University-Hospital System', in J. A. Jackson (ed.), *Professions and Professionalisation*, Cambridge University Press, 1970, pp. 111–152.
9. Recently there has been a greater autonomy between the producers of knowledge and producers of medical care, i.e., there are now permanent employment posts that involve research only. However, this does not invalidate the centrality of the production of medical care in the occupational structure.
10. See note 9.
11. Jamous and Peloille, *op. cit.*, 1970. Note 8.
12. *ibid.*, p. 112.
13. *ibid.*, p. 112.
14. J. R. Ravetz, *Scientific Knowledge and Its Social Problems*, Middlesex: Penguin University Books, 1973, p. 140.
15. Sir G. N. Clark, *A History of the Royal College of Physicians*, Vol. I, Oxford: Clarendon Press, 1964.
16. A. M. Carr-Saunders and P. A. Wilson, *The Professions*, Oxford: Clarendon Press,

1933. Berlant points out that physicians were initially concerned with abstract thought rather than medical *practice*. Ref. Berlant, *op. cit.*, 1975. Note 4, Chapter 4.

17. Carr-Saunders and Wilson, *ibid.* Also W. J. Reader *The Rise of the Professional Classes in 19th Century England*, London: Weidenfeld & Nicolson, 1966, Chapter I.

18. It should be pointed out here that there has always been ranking within the professional associations, though this is beyond the scope of the present study. The history of the RCP has been thwart with struggles between the licentiates and fellows of the College. See Clark, *op. cit.*, 1964. Note 15.

19. Sir G. N. Clark, *A History of the Royal College of Physicians*, Vol. II, Oxford: Clarendon Press, 1966, p. 909.

20. Carr-Saunders and Wilson, *op. cit.*, 1933. Note 16.

21. The surgeons and apothecaries initially belonged to medieval trade guilds, the surgeons being associated with the barbers and the apothecaries being associated with the grocers.

22. It has been argued that early science can be seen as the possession of craftsmen and traders (i.e., direct producers). See H. Braverman, *Labour and Monopoly Capital: The Degradation of Work in the 20th Century*, New York and London: Monthly Review Press, 1974.

23. Carr-Saunders and Wilson, *op. cit.*, 1933. Note 16. p. 295.

24. Although I am not suggesting that there were not scientific developments during this period. There were, but they did not necessarily permeate medical practice.

25. Carr-Saunders and Wilson, *op. cit.*, 1933. Note 16. p. 75.

26. *ibid.*

27. This point will be discussed in more detail in the next section.

28. N. Parry and J. Parry, *The Rise of the Medical Profession: A Study in Collective Social Mobility*, Croom Helm, 1976.

29. See note 24. The importance of the stratification within the Royal Colleges can be seen here. Whilst surgeon-apothecaries (doctors with licences from both the Royal College of Surgeons and the Society of Apothecaries) became general practitioners, anyone with a licence from the Society of Apothecaries was debarred from honorary hospital posts. 'Pure' surgeons (with a licence from the RCS alone) did take up voluntary hospital posts and, as we shall see below, were thus very prestigious.

30. Parry and Parry, *op. cit.*, 1976, Note 28.

31. Reader, *op. cit.*, 1966, Note 17, Chapter 3.

32. Parry and Parry, *op. cit.*, 1976, Note 28, Chapter 6.

33. Reader, *op. cit.*, 1966, Note 17, Chapter 3.

34. W. B. Walker, 'Medical Education in 19th Century Great Britain'. *Journal of Medical Education* 31. 1956, pp. 765-776. S. T. Anning, 'Provincial Medical Schools in the 19th Century', in F. N. L. Poynter (ed.) *The Evolution of Medical Education in Britain*, London: Pitman Medical Publishing Co. Ltd., 1966, pp. 121-134.

35. The most noticeable instance of such a struggle, which will not be discussed in this paper, was to culminate in the removal of general practitioners from the cottage hospitals in the first few years of the National Health Service. See R. Stevens, *Medical Practice in Modern England: The Impact of Specialisation and State Medicine*, New Haven and London: Yale University Press Chapters 8-11.

36. G. Rosen, *The Specialisation of Medicine with Particular Reference to Opthalmology*, New York: Froben Press, 1944.

37. *ibid.*
38. A few disease-entity illnesses had been recognised earlier, e.g., the plague.
39. Rosen, *op. cit.*, 1944, Note 36.
40. Stevens, *op. cit.*, 1966, Note 35, p. 16.
41. B. Abel-Smith, *The Hospitals, 1800–1948*, London: Heineman, 1964.
42. The RCP still managed to control medical practice in London, but not outside. The British Medical Association has its origins in a provincial medical society established in Worcester.
43. Reader, *op. cit.*, 1966, Note 17. Chapter 4.
44. The association between this conception of 'competition in equality' and liberal ideology is obvious. See Berlant, *op. cit.*, 1975, Note 4, for a discussion of the effect of liberal ideology on medicine.
45. Carr-Saunders and Wilson, *op. cit.*, 1933, Note 16. pp. 76–8.
46. The ability of the universities to obtain the right to give medical licenses was usually predicated on some form of consultation with the Royal Colleges over the medical curriculum. See, for example, Clark, *op. cit.*, 1966, Note 19.
47. Clark, *op. cit.*, 1966, Note 19.
48. Though the General Medical Council could be more easily described as a medical body.
49. Walker, *op. cit.*, 1956, Note 34.
50. Clark, *op. cit.*, 1966, Note 19.
51. Stevens, *op. cit.*, 1966, Note 35. p. 25.
52. G. Rosen, *op. cit.*, 1944, Note 36.
53. The idea of a 'necessary' technical division of labour refers here to the view that a development of knowledge may necessitate a division of labour based on scientific specialisms. Such a concept is suspect if it is seen as 'necessary' simply because a division of labour has occured: other factors apart from the development of knowledge and technique may have been responsible. However there is evidence that the development of knowledge in the nineteenth and twentieth centuries did necessarily produce a technical division of labour within the medical profession. For many of those who at one time abhorred specialization were forced to admit – once specialist knowledge had extended and specialists had established positions within the profession – that no one doctor could now learn all the knowledge necessary for non-specialized practice. Specialization had, therefore become necessary.
     See, for example, *Lancet* 2, 1908, pp. 479–480, Address of the President of the Section of Physiology of the BMA; *Lancet* 2, 1909, pp. 1196–1197, Introductory address on specialism, D. W. C. Hood, 'Specialism Properly Regulated is a Necessary Corollary to Increased, and Rapidly Increasing, Knowledge of Every Branch of Professional Life,' p. 1196; *Lancet* 2, 1910, pp. 1197–1198, 'Specialists and Specialism', P. S. Abraham; *Lancet* 2, 1912, pp. 398–399, 'Specialism in General and Genito-Urinal Surgery in Particular', Dr. H. Cabot; *Lancet* 1, 1913, pp. 369–375, 'Realignments in Greater Medicine: Their Effect Upon Surgery and the Influence of Surgery Upon Them'. H. Cushing; *Lancet* 2, 1930, pp. 1275–1278. 'The Danger of Specialisation', A. E. Barclay; *Lancet* 1, 1933, p. 142, 'Specialism in Surgery', C. R. G. Wakeley.
54. Another major group is the general practitioners outside the hospitals. I do not intend to deal with this conflict. However, see Stevens, *op. cit.*, 1966, Note 35., for an examination of the relationship between the development of science and the plight of the general practitioner.

55. Abel-Smith, *op. cit.*, 1964, Note 41.
56. *ibid.* There is little doubt that both these factors were a major stimulus to the creation of the special hospitals, some practitioners being primarily enthusiastic about developing knowledge, whilst others simply wanted to gain voluntary hospital posts through this means in order to establish successful private practices.
57. This caused considerable annoyance to the profession (or rather the 'elite' section of it), which was frequently expressed in the medical press. See for example: *Lancet* **2**, 1906, Leading article: 'What is Specialism?' pp. 882–883. *Lancet* **2**, 1910, Introductory address on Specialism by D. W. C. Hood. pp. 1196–1197.
58. *ibid.*
59. See note 53.
60. H. Cabot, 'Specialism in General and Genito-Urinal Surgery in Particular', *Lancet* **2**, 1912, p. 399.
61. In the field of pathology, for example, see: S. C. Dyke, 'The Organisation of Clinical Pathology to the Present Day'. in W. D. Foster, (ed.), *A Short History of Clinical Pathology*, Livingstone Ltd., 1961, pp. 124–142; P. Davey, 'The Widdicombe File: The Base Degrees by which We Rise', *Lancet* **1**, 1956, pp. 356–377; W. H. McMenemy, 'The Future Position of the Pathologist in Medicine', *Lancet* **2**, 1958, pp. 841–844; A. G. Signy, 'Trends in Clinical Pathologists', *Journal of Clinical Pathology* **23**, 1970, pp. 744–750. For a discussion of the position of pathologists, psychiatrists and radiologists, see R. Stevens, *op. cit.*, 1966, Note 35. It should be noted here that, prior to the National Health Service, 'honorary' hospital doctors did have a say in the running of the hospitals. Salaried doctors did not. Honorary doctors were supposedly unpaid, giving a charitable service, though as the years passed it became customary for the hospitals to give some reimbursement for their services. See B. Abel-Smith, *op. cit.*, 1964, Note 41.
62. In order that the voluntary hospitals should succeed in being good advertisements for the talents of the doctors, it was important that a high rate of success should be shown. Chronic cases were excluded from the voluntary hospitals. See B. Abel-Smith, *op. cit.*, 1964, Note 41.
63. In this section, the quotations given are typical of numerous others presented in the *Lancet* and the *British Medical Journal*. I have made a thorough examination of the views expressed in these journals on science, art in medicine, the scientific method and specialism from 1900 to the present. The relevant references and nuances of changing views are impossible to list here. The period examined here is the period in which statements about medicine state clearly that medicine is an *art*. By the First World War, medicine is considered both an *art and a science*, whilst after the war, medicine is considered predominantly to be a *science*. However, the dominant view of medicine as a science remains that of scientific generalism and yet another way of stating indeterminancy, experience, particularity and the irreducibility of the human subject to science (i.e., an ideology of art). The fact that most of the references given here are from the *Lancet* does not reflect a divergence of views (though there are slight differences due to the fact that the *Lancet* tends to reflect the views of hospital doctors, whereas the *British Medical Journal* represents those of general practitioners), but simply the fact that the *Lancet* gives greater detail of addresses given to medical meetings which are covered by both journals.
64. P. H. Pye-Smith, 'Medicine as a Science and Medicine as an Art', *Lancet* **2**, 1900, p. 309.

65. Mr E. Owen (surgeon) to the Canadian medical profession. *Lancet* 2, 1900, p. 817.
66. R. Fortesque Fox, 'Some Principles in the Treatment of Chronic Diseases', *Lancet* 1, 1910, pp. 1114–1115.
67. *ibid.*, p. 1115.
68. Sir Dyce Duckworth, 'Some Requirements for Modern Clinical Teaching', *Lancet* 2, 1913, p. 1449.
69. *ibid.*, p. 1450.
70. Mrs Bryant. 'Address to the London School of Medicine for Women', *Lancet* 1, 1906, p. 1022.
71. Leading article: 'What is Specialism?' *Lancet* 2, 1906, p. 883.
72. Indeed the strict control over standards of medical education resulting from the reforms of 1858 and onwards had soon ensured that medical education was confined to those from public schools. Once again social distinction became a most important factor in determining 'ability' in medical practice.
73. There are only a few examples of expressions of medicine as a science in the popular medical press of the early twentieth century..Where such views are found, they stress going *beyond experience* with science. For example, the new professor of pathology at Manchester University in 1904 says:

    > Our immediate perceptions do not carry us very far; but to investigate the facts and processes which lie beyond the reach of our immediate senses we can apply the experimental method. (Introductory address. J. Lorrain-Smith. *Lancet* 2, 1904, p. 1073.)

    And another states:

    > On the one hand she (medicine) is an empiric. She has learned to cure by . . . the 'method of trial and error'. Conquests over sickness aquired purely as a result of experience, without help either from *a priori* or from inductive reasoning . . . In her other aspect, medicine is not empiric, but a scientist. Who will refute me if I assert that medicine is as well as an art a science? Somewhere it is said that woman is the last thing that man will ever civilize. So the scientific aspect, the male-face of the two-visaged medicine, thinks of that female face, the empiric, with whom his lot is linked. He feels that his other half is the last thing science will ever render wholly rational. By dint of patient toil he improves her practice, showing her a reason every now and then (C. S. Sherrington, 'Science and Medicine in the Modern University', *British Medical Journal* 2, 1903, p. 1193).

    As man will one day civilize woman, science will control medicine!
74. Rosen, *op. cit.*, 1944, Note 36. p. 73.
    See also, for example, G. Adami, 'The Science and Art of Medicine', *Lancet* 2, 1920, pp. 732–5; H. Rolleston, 'The Clinical Laboratory in the Modern Hospital', *Lancet* 2, 1922, pp. 847–848; B. Moynihan, 'The Relation of Medicine to the Natural Sciences', *Lancet* 2, 1925, pp. 115–117.
75. Rosen, *op. cit.*, 1944, Note 36. p. 76.
76. *ibid.*, p. 77. Rosen's emphasis.
77. See Abel-Smith, *op. cit.*, 1964, Note 47, for a detailed analysis of the battles between the general practitioners outside the hospitals and hospital doctors who often gave free specialist out-patient treatment to the poor.

78. *ibid.*

79. *Lancet* **1**, 1923, 'A Scheme for a County Pathologist', p. 1278.

80. See A. Linsey, *Socialised Medicine in England and Wales: The NHS, 1948–1961*, London: University of North Carolina Press, 1962, Chapter I, and Abel-Smith, *op. cit.*, 1964, Notes 41.

81. This is if the British Medical Association was truly representative of general practitioners. See Stevens, *op. cit.*, 1966, Note 35.

82. The 'firm' is a group of doctors at various stages of training under a fully fledged hospital consultant. The trainees do much of the consultant's work (and make part-time NHS posts and private practice possible), but it is the consultant who is responsible for the patients in beds allocated to him or her. See Stevens, *op. cit.*, 1966, Note 35, particularly chapter 3, for the importance of 'ownership' of beds for authority in the hospital system.

83. Indeed Stevens points out that '. . . possibly no more than 50 or 60 consultants . . . range over vital areas of professional and thus administrative policy making', and that they are also very important in the Royal Colleges. Stevens, *op. cit.*, 1966, Note 35, p. 280.

84. Teamwork does occur on an informal basis and in some instances (in particular in special units requiring the most sophisticated techniques) on a formal basis. However responsibility for a patient with one doctor is the norm even where teamwork occurs in practice.

85. J. Powles, 'On the Limitations of Modern Medicine', *Science, Medicine and Man*, Vol. 1, Pergamon Press, 1973, pp. 13–14.

PART IV

THE SCIENTIFICATION OF TECHNOLOGY

THE 'SCIENTIFICATION' OF
TECHNOLOGY

GERNOT BÖHME
Technische Hochschule, Darmstadt
and
WOLFGANG VAN DEN DAELE and WOLFGANG KROHN
Max-Planck Institut, Starnberg

1. The Unification of Science and Technology in the Renaissance

1.1. The Emergence of the Science-Technology Relationship

In medieval times the construction of cathedrals was assigned to architectural workshops. For the organization and the techniques of construction the architects and craftsmen relied on traditional knowledge which was often kept secret by the rules of the guilds and which was only slightly modified with the acquisition of new knowledge. The construction of the largest cathedral of late medieval times, the cathedral of Milan, begun in 1386, raised unexpected problems of mathematics and statics (1). The recently acquired economic and political status of Milan called for the largest building of the period, but the city desired that the design did not follow northern European rules of construction. Within the Roman tradition of Lombardian aesthetics the northern Gothic style was considered too arching and the supporting system of pillars and flying butresses was felt to be confusing. Further, this style was determined by a well-established principle of construction which stipulated that the height of the church be equal to its width. The Milanese workshop, however, decided to work from an equilateral triangle as cross section. They hoped that the less arching elevation of this design would allow for the elimination of the confusing system of support, even if the size of the cathedral exceeded the Gothic constructions.

Northern Gothic architecture represented the most advanced knowledge and experience of the time. As the records of the Milanese

Krohn/Layton/Weingart (eds.), The Dynamics of Science and Technology.
Sociology of the Sciences, Volume II, 1978. 219-250. All Rights Reserved.
Copyright © 1978 by D. Reidel Publishing Company, Dordrecht, Holland.

workshop tell us, its members were not aware that their decision to deviate from this experience would lead them into serious problems. First, it would surpass the ability of the architects to determine arithmetically the height of an equilateral triangle. Second, even if an arithmetical determination were at hand, it would be impossible to proceed with the construction according to the Gothic rules of simple yardstick measurement. Working from the *ad triangulum* model, architects would be unable to calculate in advance the length to which workmen should cut the subsections for the main pillars. The council of the town summoned a mathematician who lived nearby. Fortunately, after some calculations he could offer a satisfactory solution (2).

The mathematical problem was followed by a problem of statics. It would be a great risk to carry out a project of this size in disregard of the Gothic techniques of supporting buttresses. The council of the town again became suspicious and called in scientific experts, this time foreigners from the north. The expert opinion of one of them, the French engineer Jean Mignot, was a scathing criticism: according to the geometric theories of Gothic statics, he observed, the whole edifice would not be solid. The theory required much stronger piers to support the system, while to make the project feasible in accordance with the geometric method it would also be necessary to elevate the main aisle. The Milanese builders and architects, he said, were not prepared to deal with the scaling-up problems arising from an extreme enlargement of the design of Lombardian provincial churches. Mignot's opinion regarding their ideas on statics is recorded:

Proposals were made in a fashion more willful than sound; and what is worse, it was objected that a science of geometry should not have a place in these matters since science is one thing and art another. Said Master Jean [Mignot] said that art without science is nothing (*ars sine scientia nihil est*) (3).

In opposition, the local master-builders appealed to the Lombardian tradition of architecture. They claimed that they would not find insuperable difficulties in transposing Lombardian techniques of support to more monumental and vast projects. The practice-oriented reply to Mignot ended with the statement, '*scientia sine arte nihil est*'. The local guilds won their argument with the foreign experts, although their concepts were not well-supported either theoretically or by experience. The only argument which speaks in favor of the concept of *'scientia sine arte nihil est'* is that although it was built contrary to the Gothic theory, the cathedral of Milan still stands (4).

This dispute between the experienced practitioners and the calculating theoreticians presupposes that, in principle, there exist two different approaches to solving the same problem. This is the first time that science and technology can be seen to compete in being respectively the better means to fulfilling a certain end (5). From the Greeks down to the early Renaissance, the tension of the relation between theory and practice was primarily related to the moral question of which life style – the theoretical (contemplative) or the life of labor – was more delightful in the eyes of God. In the Milan controversy neither life styles nor world views are questioned but rather the practical utility of theoretical knowledge and the deficiencies of traditional knowledge for solving new problems. The spokesman of the faction which supported theory was not a philosopher but an engineer. And it was to be engineers, artists and mathematical practitioners who were to play a decisive role in the development and social acceptance of this new type of utility-oriented theory.

1.2. Early Technical and Natural Sciences

These new theories – or at least new theoretical claims – cannot unequivocally be classified, either under the rubric of the arts or under that of the sciences. This is only in part due to the fact that in the Renaissance period *scientia* and *ars* meant something different than what they mean in modern times: it is just as difficult to assign these theories to the by now familiar subdivisions of the engineering sciences and the fundamental sciences. The reason for this ambiguity is rather that these theories violate the traditional boundary between the natural and the artificial, between understanding phenomenona (theoretical consideration) and constructing artefacts (poietic practice).

Leonardo da Vinci (1452–1519) was one of the first scientists to combine the technical construction of a desired reality with the discovery of the structures of a given reality. Thus Leonardo came upon new ideas in hydrophysics both by solving problems which had arisen with the new projects for regulating water-flows (such as construction of canals, regulation of rivers, artificial irrigation), and by his observations of meteorological or hydrological processes in nature (for instance the formation of clouds and water whirls) (6). Leonardo's discoveries were as much laws of nature (*ragioni*) as rules of operation (*regole*) (7).

In the mid-sixteenth century, Nicole Tartaglia (1499?–1577), a mathematician as well as an engineer, was working on a theory of ballistics which combined a natural force (gravity) with an artificial (the impulse of the projectile) and led to a unified geometrical representation for both. Tartaglia published his theory in a book which bore the programmatic title, 'Nova Scientia'. It was addressed to "any speculative mathematician, artillerist and others". It would be difficult to decide whether this book deals with technical or natural science (8).

The association of natural philosophy and technology and the transformation of both into a new type of science occurs in the fifteenth and sixteenth centuries in many fields of the then relevant sciences. Paracelsus (1494–1541), Ambroise Paré (1510–1590), and Andreas Vesalius (1514–1565) revolutionized the human sciences based on natural philosophy as well as on the medieval art of medicine by laying the foundations of the disciplines of pharmacology and surgery (9). William Gilbert (1544–1603) framed a theory of magnetism intended to encompass earth magnetism and artificially induced ferromagnetism as well (10). Peter Apian (1501–1552), Gerard Mercator (1512–1594), and others worked on theories and processes intended to coordinate terrestrial orientation and astronomy (11).

1.3. The Knowledge of Possible Nature – The Knowledge of the Nature of the Possible

Within the science of the late Renaissance there are innumerable examples of scientists engaged in connecting science and technology. One may very well ask whether these attempts were accompanied by the development of a new conception of nature, interrelating its explanation and control. The scientists and philosophers claimed, at least, to have done this.

In his 'Discourse on Method', a later and thus a more reflexive exposition than those of the Renaissance scholars, Descartes

(1596–1650) gave account of this new concept of nature and of natural knowledge:

But as soon as I had acquired some general notions concerning Physics and as, beginning to make use of them in various special difficulties, I observed to what point they might lead us and how much they differ from the principles of which we have made use up to the present time, I believed that I could not keep them concealed without greatly sinning against the law. . . . For they caused me to see that it is possible to attain knowledge which is very useful in life, and that instead of that speculative philosophy which is taught in the Schools, we may find a practical philosophy by means of which, knowing the force and the action of fire, water, air, the stars, heavens and all other bodies that environ us as distinctly as we know the different crafts of our artisans, we can in the same way employ them in all those uses to which they are adapted, and thus render ourselves the masters and possessors of nature. This is not merely to be desired with a view to the invention of an infinity of arts and crafts . . . but also principally because it brings about the preservation of health . . . (12).

To achieve these goals, it is necessary to discover "the principles, or first causes of all that is or can be in the world" (13). Recasting nature to something that is conceived of being possible but does not exist as yet – this is technology. If, therefore, the concept of nature is not confined to that which is objectively given, but covers that which is objectively possible, the concept of nature will include the procedures and products of technology. The purpose of natural knowledge is not only to discover the facts but also to construct 'arte-facts' according to rules which delineate the realm of the possible states of nature. In the context of medieval craftsmanship, rules were simply instructions for making something. For Descartes they have become simultaneously both the laws which were put into nature and also those which establish its system of possible operations (14). Some notes of Leonardo indicate that this concept of nature already underlay the investigations he made some one hundred years earlier. About the utility of his results, Leonardo writes:

These rules enable you to discern the true from the false, and thus to set before yourselves only *things possible* . . . (15).

If you were to ask me what do your rules bring about and of what use are they? I should answer you that they prevent the inventors and investigators from promising themselves and unto others *things that are impossible* . . . (16).

Finally, a consideration of Leonardo regarding the concept of nature as the reality which can be constructed with some particular end in view:

> You who speculate on the nature of things, I praise you for not knowing the processes which nature ordinarily effects of herself, but rejoice if so be that you know the issue of such things as your mind conceives (17).

With this conception the phenomenalism of Aristotelian natural knowledge is replaced by a constructive point of view: *knowledge* of nature has become identical with its experimental and deductive *construction*.

Cassirer has described the new thought underlying this program as the transition from substantive to relational thinking, a process which received its main impulse from the artists and engineers of the Renaissance. The parallelism of the artistic formation of nature and its geometrical description gave primacy to the formal regularities in nature rather than to its ontological qualities (18).

Piaget characterizes this transition from substantive to relational thought as a decentering process in which it is learned that the framework through which we perceive nature is constructed according to cognitive operations (19). Kant stated that this revolution of our 'Denkungsart' was brought about by the 'realization' [Einfall] "that we only learn from nature, what our mind superimposes on it" (20). This nexus of cognition and operation provides a more profound insight into the interdependency of the knowing subject and the objects to be known than did either the ideas of traditional natural philosophy or those of advanced craftsmanship alone.

1.4. The Historical Evolution of the Interrelation of Technology and Science

Historically, this unity has been built up only step by step and only in particular sciences and technical fields. On the one hand, this has been due to the fact that the social acceptance of the new sciences depended on external chances of selection which arose only gradually as capitalist society developed. Further, the application to the different areas of nature of the epistemic principle that knowing and operation are interdependent was hampered by empirical problems which could only be resolved successively.

In what follows the process of the intellectual realization and social acceptance of the program of the new sciences is subdivided into three phases:

The first phase (ca. 1660–1750) begins in the Period of the Restoration in England and the spread of Absolutism on the Continent. From the vantage point of the history of science it is marked by an institutional and cognitive differentiation of the areas of science and technology. Nonetheless, scientific orientation to techniques remains important in two respects. First, it gives rise to a special development of technology, i.e., the technology of scientific instruments and procedures. Secondly, the principle that to know something means to know how it operates is worked out at a universal pattern of explanation in the form of the mechanistic image of the world.

The second phase starts with the industrial revolution and approximately covers the nineteenth century. The spread of capitalism and its internal dynamics makes technical invention a constituent element of economic reproduction. The dynamic momentum given to technology produces a demand for science, bringing about the first examples of a process that may be called the 'scientification' of technology. At the same time a wider scope is given to the economic utilization of those technologies which were originally developed as scientific instruments and procedures.

In the third phase, the reciprocity in demands and supplies between science and technologies becomes systemized and strategically planned. This process begins in the latter half of the nineteenth century and comes to play an increasingly significant role in science policy and planning of research in the twentieth century. Scientific advance becomes goal-oriented and technologies are planned in conformity with theories.

2. The Separate Development of Natural Science and Technology

Given the background of the first section, it is not surprising that the institutionalization of the 'new sciences' in the seventeenth century comprised both their tradition of natural philosophy (cosmology, theory of matter and force, etc.) and their technological traditions (mechanics, time keeping, navigation, mining, etc.). The charters of most of the scientific institutions following the foundation of the Royal Society and the *Académie des Sciences* propose scientific research which is to lead to results both new and useful.

It is, however, from a sociological point of view, surprising that the institutionalized juncture of novelty and usefulness did not last long (21). Admittedly, innovation in industry and social structure was by no means the foremost goal of the absolutist governments, but still, innovation played a fairly important role in the domains of warfare, foreign trade, luxury production, standardization and measurement (as a basis for taxation), and transportation (22).

The conjunction of book learning and craft techniques in the Renaissance had arisen from the breaking down of cognitive and institutional barriers and had been brought about by artists, engineers and individual scholars. Yet the result was not a unified 'technical science', but rather on the one hand, a new technology incorporating scientific method, and on the other, a new natural science which analyzed nature by means of technical apparatus and interpreted it according to the model of such apparatus. Though the ideas of the scientists were intimately tied to technical problems, they themselves were not guided by a 'technical interest in knowledge' (23). Instead, notwithstanding the proximity of both fields, different goals of investigation brought about a new separation of science from technology. While the engineer designed apparatus and improved their functions as the technical realization of a practical intent, the scientist tried to understand how it functioned, he was directed towards theory. Although scientists often enough began their research in connection with technology, as the investigations proceeded these took on a life of their own, independent of their technical starting points. The scientists' concern was "the true natural philosophy" (24).

The technology of pumps, for example, concerned many scientists in the seventeenth century. The phenomenon that vacuum pumps only lift water to a certain height and its explanation by Viviani that the atmospheric pressure impedes to lift it higher led Toricelli to the construction of an instrument for the measurement of the weight of the air, the barometer. This instrument, then, was used by Pascal to test the existence of a vacuum – and the ultimate result was a general theory of pneumatics.

Similar, the trajectories of projectiles were no doubt an extremely interesting subject for scientists occupied with the problems of mechanics, yet their main concern was the theory of impetus, that is, the question of how a body is kept in motion without a moving agent. In this case, theoretical interest gave rise to a development in the field of mechanics independent of developments in the field of artillery.

The scientific interest in understanding natural and artificial events acted as a source of internal problem generation which made science autonomous of production-oriented technology. Already at the turn of the seventeenth century interest in theory had come to outweigh interest in research oriented to practical needs, even in the Royal Society and in the *Academie des Sciences*, whose founding charters had stipulated that their business was to promote the trades and the crafts by means of the new science. With the Royal Society this change can be traced to the time when Newton became a Fellow (25).

As scientists came to act independently in their focus on the developments of theories and as their contributions to advances in production, warfare and navigation became fewer, the systematic improvement of these fields devolved to the engineer and engineering became institutionalized as well. The absolutist states organized engineering strictly to meet the needs of transportation, mining, the military, and the navy. This led to the establishment of the various technical corps and schools in France, the *École des Ponts et Chaussées* (1750, the *Corps des Ponts et Chaussées* had been established earlier, in 1720), the *École du Corps des Ingénieurs des Mines* (1778), the *École Royale Militaire* (1753), the *École du Corps Royal du Génie* (1665, for the education of architects), the artillery schools and the naval colleges. In England similar institutions were the Mechanics Institutes, and in Germany the Mining Academies in Berlin (1775), in Freiberg (1765), and the Schools of Agronomy. These schools trained engineers and their curriculum comprised the fundamentals of mathematics and the natural sciences. They were not always the centers of the new sciences, yet, it was within this frame that a turning point occurred which marked a rapprochment with the institutions of science in the nineteenth century.

3. The Orientation of Science Toward Technology

Although from the seventeenth to the nineteenth century the new sciences had developed both cognitively and institutionally apart from technology, they nonetheless remained technology-oriented. One reason for this seems to be that for modern science the experience of nature is mediated by instruments which are perfected in conformity with scientific standards. These standards are set up without reference to the usefulness of these instruments apart from their scientific purposes. Another primary reason for the lasting technology-orientation of science is that theory development initially evolved within the framework of the mechanistic worldview. The theories of natural science are instrumental theories, partly because they emanate directly from analyses of apparatus, and partly because their model for the interpretation of nature is the 'big clock'. Thus the internal dynamics of scientific progress involves a technical dynamic (26).

3.1. Technological Developments Internal to Science

The modern experience of nature is an experience of regularities and of apparatus as the models of nature. The experience produced according to specific rules, with data reproducible by anyone who observes these rules, is more significant than the experience of events perceived individually. The data for science is not that given directly to the senses but is its objectified occurrence, that is, its effect through and on apparatus. Both reproduction and objectification of events were already established by craftsmen and engineers; in connection with science, however, they undergo a greater degree of specification and standardization. And it is this which determines the development of scientific instruments and procedures (27).

Scientific instruments were either designed for scientific purposes as such or, where they already existed in connection with technical-practical needs, they were further developed for scientific purposes and in conformity with scientific standards. The last was the case, for instance, with the clock and with the pump. The development of the clock in its uses internal to science became dissociated from its function in the presentation of the celestial movements and was determined by the idea of the absolute isochrony of periodic processes. This was done by introducing the pendulum (28). The development of the pump followed a similar pattern. Naturally, pumps had already been employed for technical and practical purposes. But their perfection in the form of the vacuum pump sprang from the scientific interest in the phenomena revealed by the barometer, and was governed by the idea of the perfect vacuum. Philosophically, space without any matter seemed to be impossible; experimentally it was approximately feasible.

A concise description of the scientific standards that actuate the development of apparatus and new procedures is facilitated by subdividing them through a distinction borrowed from chemistry into *analytic* and *synthetic* apparatus and procedures. The term analytic refers to instruments of measurement such as the clock, the thermometer, the calorimeter, the barometer, the voltmeter and the amperemeter, and to processes such as Liebig's analysis of the elements or the detection of grape sugar by Fehling's solution. The term synthetic refers to apparatus such as the vacuum pump, the Voltaic cell and the electromagnet, and to processes such as the synthetic production of alizarin. In general, any experimental device which serves to manifest an effect or a substance can be considered such a synthetic apparatus or process.

The scientific standards guiding the development of the *analytic instruments* are the increase of *precision*, the *unequivocal* relation of data to variables, the *universal comparability* of weights and measures, and the *generalization* of measurements over the whole range of a parametric scale. Increase in the *precision* of scientific instruments of measurement by far exceeded what was practical for production technology or what the engineers could calculate mathematically (29). It is even more significant that the standards of precision became separated from the standards given by the original paradigm of nature. Scientific standards of precision were defined in accordance with theoretical ideals. Thus, even in antiquity the technical standards defining the vertical and the horizontal were transcended by Euclidian geometry. In the same way, the standards of celestial motions for the construction of clocks (the sun's daily zenith, the lunar month, the solar year) were replaced by the ideal of absolute isochrony. Thermometrics at first defined its scales using the hottest and coldest days of the year (Fahrenheit) or the states of liquid water (Celsius). These scales in turn were replaced by Kelvin's absolute scale, through which it became possible to indicate the adequacy of any of the substances previously employed to function as a temperature standard.

The requirement that instruments give *unequivocal reading* was also unrelated to technical purposes outside of science. The farmer, the metallurgist and the brewer needed indicators of those changes occurring in the processes they were working with, but as a rule they were guided by impressions which from the vantage point of science were produced by many interacting variables. Scientific instruments, on the other hand, are intended to respond to one and only one variable. Thus, after the invention of the thermometer by Galileo, Drebbel, Fludd and Santorio, other scientists tried to make it as independent as possible of air pressure and the material qualities of the substances used (30).

Until the emergence of mass production and world-wide markets, the *universality* of data was not of great practical value. To be sure, attempts to unify weights and measures had been made in various countries throughout the eighteenth century, but not until the nineteenth century were general conventions on weights and measures established. For science, on the other hand, the universal comparability of data was a standard from the very outset. Nor can universal comparability be attained simply by international agreements on units of measurement, as its realization also implies overcoming a variety of technical problems (31).

In the same way, *generalization* is not a standard of immediate technical interest. Generalization means the establishment of procedures of measurement covering the whole range of a parametric scale. The potter has one method for determining temperature, and the beerbrewer a different one. Correlation of their scales is a matter of indifference to both of them. Scientists, however, search for a theoretical concept of heat for which the different procedures of measurement are but partial interpretations.

When in the nineteenth century production started making severe demands on purity of materials, precision of machine parts, and universality of measures and weights all over the world market, the analytic apparatus and procedures for testing materials and the control of processes were available as a result of a development internal to science and could be modified for industrial purposes.

The development of *synthetic apparatus and procedures* are guided by different standards. Empirical investigations within the framework of modern natural science thus presuppose that phenomena can be isolated, are reproducible, and are disposable regardless of regional or seasonal factors. Thus, though electricity is made manifest by rubbing amber against a cat's fur or by a flash of lightning, for the advancement of the science it was important to isolate electrical phenomena from related ones like magnetism and to make electricity freely available, as was accomplished by Hauksbee's Electrical Machine, and later by the Leyden jar and Volta's electric pile. Analogous examples are to be found in chemistry. A condition for analyzing the structures and properties of a substance is to isolate it, and to produce it in sufficient quantity and purity. Synthetic apparatus and procedures serve to isolate and present substances and effects, and where they exist only in insufficient strength or volume for investigation, to store, accumulate and reinforce them.

When a need for these substances and effects emerges in the economy or in the industry of warfare, techniques developed within science itself become a part of production technology.

3.2. The Technology-Orientation in the Theory of Mechanics and in the Mechanistic World Image

Science and technology did not only diverge in the development of apparatus and procedures; they also differed in their theoretical concepts. Even though mechanistic theories were supposed to cover, at least 'in principle', technical effects, in their formation it was necessary to develop concepts which had no bearing on technological purposes extraneous to science. The most conspicuous example is the idealization of bodies to 'mass points' in mechanics after Newton. The formation of scientific theories took a different direction in reference to the ideal of producing useful knowledge. Since it was either mathematically too complex or not worth its efforts to 'deduce' technical devices from mechanistic theories, little contribution could be made to technological advance.

In the period of the Renaissance, technology and theoretical knowledge were still connected only on the basis of Aristotelian categories. In interpreting their discoveries, the practitioners of the experimental method still had to appeal to the 'natural place' of bodies and to teleological forces moving them. The unity of technology and science was practiced in experimental research, but its theoretical interpretation lacked consistency. Only in Galilei's new physics was it possible to conceive physical (natural) movements as technical (artificial) movements and vice versa.

Descartes' mechanistic philosophy realized this unity, although his theories in physics and biology were rationalistic constructs rather than explanatory frameworks guided by experience (32). Descartes' model of nature was a technical one – change of motion is caused solely by push, pull and pressure – i.e., he explained natural by artificial motion presupposing the conservation of a constant amount of motion. In this way, the forces which were known as operative in mechanistic devices were held to be the only possible forces existing in nature. But, even disregarding its inherent difficulties, this argument does not provide the whole answer. To explain nature from the principles of technical mechanics ultimately requires that mechanics be explained from nature.

If mechanics is supposed to be not only a simulation model of nature, but also its real representation, then the mechanical forces must be derived from some universal principles of movement as such. Newton's theory of motion and force was highly successful in formulating this common basis of science and technology (33). Nonetheless, a comparison of this scientific mechanics with the technical mechanics of the so-called five simple machines–wheel and axle, the lever, the pulley, the wedge, and the screw – shows that although Newtonian mechanics explains the functioning of these machines, the machines themselves no longer figure in it. The mathematical derivation from the Newtonian principles entails nontrivial processes of integration. For this reason, the engineers continued working with the 'already integrated' simple machines. This relationship is found also in later fields of research originating within the frame of mechanistic approaches as, for instance, was the case with electrical engineering and Maxwell's electromagnetic theory of light. The technological study and use of condensers, resistors, valves, and so on was on the whole independent from their theoretical foundation. However, although the mechanistic theories did not lead to useful applications, their formation was oriented toward the explanation of technological problems like torsion, elasticity, fluid resistence. Moreover, many of the early theories of modern science came into existence as theories of specific instruments. This is true in physical optics with the theory of the microscope and the telescope, in pneumatics with the theory of the pump and of the barometer, and in thermodynamics with the theory of the steam engine. Theories explaining instruments enable one to distinguish contingent from systematic limitations of the instrument's performance and efficiency. This distinction is then in many cases the condition for using apparatus economically or for the construction of technical alternatives to them.

3.3. Scientific Supply to Production

The development of technology internal to science in late seventeenth and eighteenth centuries consisted in the perfecting of instruments and procedures and in their theoretical explanation. This 'scientization' of technology was not directly associated with the sphere of production nor did there exist a considerable demand for scientific technology in production.

However, given the economic dynamics which obtained after the industrial revolution in the nineteenth century, science and productive technology came closer together again. With increasing economic opportunities for technological inventions, the technological achievements internal to science became relevant to industry. Furthermore, scientists started devising economically exploitable technologies and experimental inquiry was no longer exclusively devoted to such goals as the construction of ideal regularity, the identification of elementary components or the testing of fundamental principles.

In the nineteenth century, the scientific supply to production consisted primarily of the synthetic procedures and instruments originally developed for scientific purposes. The development of the telegraph may serve as an illustration. It was based on scientific procedures and instruments which were not modelled on the techniques of craftsmanship, but which led to effects which technical-practical strategies could simply not have produced.

In 1820 the Danish professor of physics, Christian Oerstedt, discovered the magnetic effect of currents. The technical potential of this discovery for the transmission of signals was recognized immediately and the role of scientists in the development of the electromagnetic telegraph was considerable. Contributions were made by Schilling, a Russian State Councillor, and by the physicists Henry (Princeton), Jacobi (Petersburg), Gauss and Weber (Göttingen). In 1837 Cooke and Weatstone built the first practicable telegraph, which was put into operation by the London and Birmingham Railway. Cooke himself had not received a scientific education. He made a living from the sale of self-constructed anatomic models. When he started his work on telegraphs, he encountered electromagnetic problems, and to solve them joined forces with Weatstone, who was then a professor at King's college in London. Electric telegraphy was the first of a vast series of new technologies resulting from the direct exploitation of the principles and scientific discoveries of electricity. Other examples of this kind were Marconi's wireless telegraph, which was derived from the results of the work of Maxwell and Hertz (34), and Bell's telephone, which proceeded not only from the foundations of electromagnetic induction as laid by Oerstedt, Lenz and Faraday but also from Helmholtz' fundamental studies of physiological acoustics. In all these cases science had already assumed the role of supplier to production.

4. The Orientation of Technology Toward Science

So far we have analyzed the orientation toward technology inherent in the development of the theoretical sciences. We shall now examine tendencies to reconcile science and technology which were fostered by advances in those technologies not tied to scientific purposes. The technological advances central to the scientific revolution were developed without recourse to scientific theories and without significant transfer of the experimental results of academic or university science. This is true of early labor saving machinery (the loom), power engines (improved hydraulic motor), transport (the locomotive), and metallurgy (cast iron).

The relationship between science and technology was visible only in their partially congruent methods of research. Experimental variation, measurement and quantification, and geometric description had become procedures which promoted the development of technology. Many of the achievements of the industrial revolution stem from such a method-ological development of technology (35). On the other hand, under Absolutism, governmental regulations as well as the guild's control of production inhibited innovation and consequently slowed down the rate of technological progress relative to the special development of scientific instruments. The industrial revolution, brought about by a few innova-tions with key significance, in turn set new incentives for technological research and established a demand for scientific knowledge and methods in technology.

The following four points indicate the growing need for technological help in industry, which led to a demand for science:

1. *Additional raw materials.* In many fields a shortage of raw materials limited industrial expansion. The availability of saltpeter, soda, and steel determined the volume of technology-related outputs in the chemical and the engineering industries. Efforts were made to overcome these limits by means of scientific analysis, frequently through attempts at the synthetic production of the respective raw materials.

2. *Increasing efficiency.* Low efficiency can render a technology economically useless. Thus the early Newcomen steam engine, because it consumed enormous quantities of coal, could only be used in mining areas. The same applied to the locomotives used at the start of the nineteenth century, which only had a speed of a few kilometers per hour and thus could not be used for other purposes than for transport in mining. Optimizing the efficiency of the steam engine became an object of science. Carnot, the French engineer and physicist, theoretically established the principle of the thermal engine and projected the maximum efficiency achievable in each case.

3. *Scaling-up of critical orders of technical dimensions.* The increase in the dimensions, the mass, the pressures, the temperatures, and the like, of technical procedures led to mainly non-linear changes in stability conditions, in quality of materials, and in modes of operation, which could no longer be achieved on the basis of existing technology.

4. *Greater precision of technical procedures.* Precision comprises guarantees of the purity of products and materials (thus the availability of pure yeast for proper fermentation was a major problem for brewery

techniques) of the accuracy and constancy of the size and material qualities of machine parts, and finally, of the accurate control of industrial processes and the correction of disturbing effects.

These problems cannot always be resolved through step by step improvements on existing procedures. The substitution for natural of synthetic fertilizers, the development of a separate condenser for the steam engine, the replacement of the bucket-wheel by the turbine in the use of water power, and the like, often start with or even require the adoption of principles other than those which underlie the established procedures. If social conditions are static, the limitations of improving traditional techniques often constitute the limits of technological progress as such. Technological progress exhausts itself in the slow perfecting of already known principles, as with the water wheel and the windmill. Only under dynamic social conditions do these limits constitute a challenge and generate a demand for science.

To develop technology beyond these limits, scientific analysis is, if not a necessary, at any rate in the long run a superior strategy. On the one hand, theory provides technical principles which cannot be discovered on the basis of extrapolations from known processes. Theory can play the role of an heuristics of invention (36). On the other hand, scientific research can open the 'black box' of functional interrelations by subjecting the factors involved to causal and elementary analyses. In this way, both the theoretical optimum of a technical process, and the limiting conditions of the materials and procedures actually applied, become determinable. Thus, theory offers technical prognoses, in some cases even for non-linear modifications to which the rule-of-thumb method is inapplicable.

Two examples may illustrate how scientific analysis served to solve problems of technological progress. The first instance is the improvement of steel production by the Gilchrist–Thomas Basic Process (1871–1879), invented by the cousins Percy Gilchrist and Sidney Gilchrist Thomas. In 1856 Bessemer's converter had signalled the decisive technical break to cheaper and speedier production of steel. Bessemer's process, however, had failed to remove the phosphorous which most English and European ores contained. The problem of making steel from phosphoric ores was tackled by Thomas. He had worked as a junior police court clerk in the London docks and continued his studies at Birbeck College (a kind of 'open university' for artisans and others in full-time employment) in the evenings, concentrating on chemistry. He read the scientific and technical literature, and after four years of scientific analysis and experimenting he found the solution of the problem: it was that of absorbing the oxidized phosphorous or phosphoric acid in a basic lining made from magnesium limestone (37).

The second example is that of the "laws of the construction of locomotives" by Ferdinand Redtenbacher (1855). The building of the locomotive and the working out of construction problems, associated primarily with Robert Stephenson, who insured that the engine's weight rested equally on all wheels so that they would not spin and developed a direct drive between the cylinder piston and the driving wheels, was not the result of any notable contribution on the part of science. In the introduction of his 'laws', Redtenbacher stated that practice "with its sound instincts and perceptions" had reached a stage close to perfection (p. III). As with other major constructional engineering techniques, all that was left for science to do was to keep running after the technical dynamic with theoretical explanations in hand. But there was a point where the building of locomotives was connected with science, i.e., in the problem of the steadiness of locomotives at high speed. The unbalanced reciprocating weights of pistons, crossheads and connecting rods, which were originally set to cancel out only straight fore-and-aft forces, at high speeds led to dangerous oscillations which could accidentally throw the locomotive off the tracks (38). The engineers tried to improve matters through a more equal distribution of the weights on the locomotives (e.g., by moving the cylinders rearwards), but it was difficult to design the dimensions and the setting of the reciprocating weights. Redtenbacher, who was a Professor at the *Polytechnische Schule* in Karlsruhe and one of the founders of scientific constructional engineering, attempted a fundamental solution of the problem by means of a theory of disturbing effects in express locomotives. He defined the various forms of the disturbing effects –jerking, rolling, nosing, surging, pitching – and the forces causing them. He worked out a set of differential equations determining these movements and stated the methods for solving them, and further defined both the conditions and the limiting factors for eliminating the disturbing effects of the imbalance of reciprocating weights. It is not certain whether

Redtenbacher's formulas came to have any importance for practice. Possibly it was only the theory of balancing of le Chatelier (1849), who was more closely associated with locomotive building than was Redtenbacher, which actually came into application. Nonetheless, it was a *theory* of the disturbing effects which ultimately resolved the problem of dynamic balancing by making it possible to predict with some confidence the results of the said effects. This theory is even today an essential element in mechanical engineering.

'Applied science', theoretical and experimental research aiming to solve given technical problems, is a characteristic form of the interaction between scientific and technological development in the nineteenth century. On the one hand, this interaction was rooted in the demands for technical advances prompted by the industrial dynamic of the age, and on the other, in the increasing potentialities and proficiency of scientific output (39).

In the writings of the scientists of the nineteenth century, the growing significance of science for technology is frequently seen not as the result of a *demand* from technology for scientific problem-solving, but rather as the result of the *supply* of new techniques following upon the autonomous development of theoretical science. The industrial expansion experienced by society indeed had the effect that in the nineteenth century the technical potential of scientific discoveries was increasingly applied in technology. We have illustrated this point by the telegraph and other techniques in the field of electricity which exemplify a technical dynamic made possible by offers from science. This gave rise to the general argument that pure or fundamental research is the basis of the technical dynamic. With this argument the scientists sought to protect their recently gained professional autonomy against the demand for a science oriented toward technical applications. Pasteur said that there was no category of science which could be called an 'applied science'. "There is science and there are the applications of science joined together as is the fruit to the tree on which it grows" (40). Yet Pasteur himself had spent many years of his life in the search for scientific solutions to special practical problems, and his central preoccupation was always with the discovery of the causes of a problem and the scientific explanation of the factors involved.

Except for the ideological reason just mentioned, it is hard to understand why this type of externally oriented research directed at the medical-technical solution of urgent problems should not bear the name of applied science. Pasteur's medically-oriented microbiological research, and similarly Liebig's agricultural chemistry, are both closely related to the 'demand model' of science-oriented technology which has been exemplified by the Thomas converter and the theory of locomotives.

5. Theoretical Technology

The relationship between scientific progress and technological innovations in the nineteenth century was still unsystematic. This applies both in the case of scientific inputs into new technologies and in that of the demand for scientific solutions to given technical problems. The relationship between science and technology was determined by contingent factors such as personal contacts among scientists and practitioners, the technical and economic interests motivating individual scientists, and the degree to which technicians had had access to scientific education (41).

At the end of the nineteenth century and increasingly in the twentieth century the technical dynamic actuated by capitalism and by war leads to an expansion of industrial research and development, to independent institutions in applied research, and to a shift of technical education from training at the shop to academic learning. The connection between science and technological progress is in many aspects an offspring of the nineteenth century 'supply model'. New technologies emerge as the 'fall-out' of basic research. The structure which characterizes the new level of the interaction between science and technology in the twentieth century, however, is the ramification of science into special technological theories, or, from the vantage point of technology, the pursuit of technical goals through the construction of theories.

Natural knowledge and technology are no longer only methodologically equivalent, nor are they linked only by the epistemic structure of operational knowledge. They now tend towards unification at the level of theory. In the twentieth century the development of useful technology through scientific theory building becomes possible and can be planned strategically. Its basis is the successful formulation of theories for a series of object areas. In contrast to the conditions from which science started in the nineteenth century, science at present disposes not only of successful theories of mechanics, but also of reliable electrodynamic and chemical theories: for such phenomena as moving bodies, heat, light, electricity, science has achieved solutions of most of the fundamental problems, and these solutions can serve as the foundation of special theoretical models, explaining complex technical phenomena. The process of interconnecting technical phenomena with a fundamental theory by means of special models has two complementary aspects. It includes both the formulation of theories for technical structures and the concretization of general scientific theories. The examples of the sciences of chemical engineering and fluid mechanics will serve to illustrate both these aspects.

Toward the end of the nineteenth century the engineering design and optimization of chemical-technical production processes were in the main still based on personal experience and trial-and-error variations. By the mid-twentieth century, however, they are the subject matter of a theoretical science: the science of chemical engineering (42). The development of chemical engineering as a scientific discipline covers three distinct stages. The first, which lasted until World War I, is marked by the aggregation of the disciplinary competencies of the engineer and the chemist in the area of chemical engineering. (In Germany this took place through the cooperation of traditionally trained engineers and chemists, and in the U.S.A. and Great Britain through the formation of the profession of the 'chemical engineer'.) New processes were increasingly being developed experimentally in engineering research laboratories. But translating the results of this research into practice was still a matter of chance and 'finger knowledge'. The disciplinary methods of chemistry and engineering were not yet fused, and therefore a direct scientific treatment of the complex operations in the 'real' technical processes was not feasible.

The second stage of 'scientization' – especially in the U.S.A. after World War I – consists in classifying production processes on the basis of unit operations. The units of technical operations such as distillation, filtration, mixing, crushing, grinding, and the like were differentiated from each other and organized according to practical requirements. The operations were dealt with by means of scientific methods, but these methods were not directed towards theoretical analysis on the level of fundamental sciences such as molecular chemistry, thermodynamics or hydrodynamics. By about 1930, chemical engineering had been established as a scientific discipline with its own methods, standard solutions and conception within the framework of phenomenological theory.

The third stage, beginning in the thirties and accelerating after the Second World War, entails the integration of chemical engineering into the framework of physical and chemical theories. The characteristic of this third stage is that the real processes in chemical reactors are more and more successfully made comprehensible by means of microtheoretical concepts within the framework of thermodynamics, hydrodynamics, aerodynamics, kinetics, and other theories. Work on chemical engineering problems becomes for scientists a possible avenue for the continuation of physico-chemical research (43). They develop special concepts and mathematical models which interconnect the more complex technical processes and operations ('macrophenomena') with the idealized objects of fundamental science. Thus, for example, 'macrokinetics' is developed to understand chemical reactions which are influenced by physical transport phenomena (transport of heat, energy and matter).

In the ideal case the theoretical treatment of chemical engineering makes possible the theoretical calculation of the conditions of technical apparatus and of process operations. It also allows engineers to design productive means with a minimum of uncertain extrapolation on the basis of existing processes or trial-and-error experimentation.

The second example of a 'theoretical technology' is fluid mechanics, with its various special models such as that of the friction of bearings and of lubrication, or that contained in the theory of the airfoil (44). As was the case with chemical engineering in the third stage of its development, fluid mechanics is not only 'in principle' related to specific technical devices, (as was Newtonian theory, for instance, to the windmill) but its special theoretical models also instruct the engineer how to construct a particular device: they make possible the calculation of the required technical solution, though its feasibility may remain a problem. These theories make it possible, by varying certain parameters and holding others constant, to select optimal solutions. They also predict what will occur if thresholds of a certain order of magnitude are surpassed, and they identify the causes of the limitations to which a particular technique is subject.

As opposed to chemical engineering, theories of fluid dynamics may be

called theoretical technologies: they did not emerge from the systematization of the corresponding technologies. In the nineteenth century
theoretical and technical hydraulics evolved in parallel with almost no
connection to each other. This was due to the fact that the Navier-Stokes'
equations, which 'in principle' had made classical hydrodynamics
encompass all the processes of fluid mechanics, were solvable only in a very
few instances which had a bearing on technology. The transition to
technological theories in the field of the mechanical effects of fluids was
the result not of systematizing technology, but of theoretical efforts. It was
Prandtl's concept of the boundary layer which enabled the transition to
the 'real' technical processes to be made. To put it briefly, the boundary
layer concept allowed fluids on the whole to be treated as viscosity-free,
but required that regard be paid to the effects of viscosity at the boundary
layer. The boundary layer concept made it possible to specify general
hydrodynamic theory into distinct theories, which covered particular
types of technical operation such as lubrication, the airfoil, and the
propeller.

The technology of aeronautics was from the start a scientific
technology, not only in the methodological sense that the first designers of
aircraft made scientific experiments (the brothers Wright even had a wind
tunnel) but also in the sense that theory played a crucial role in the development of this technology. Kutta's theory of the airfoil wing, by virtue of
which one could calculate the lift of the wing, under idealized conditions,
did not have any practical bearing on the first flights made. But Prandtl's
theory of the airfoil for real aircraft wings (which, unlike Kutta's, were of
'finite length') made the theory practicable. Step by step, on the basis of
this theory it became possible to compute the lift-to-resistance ratio for
airfoil wings, to construct wings with a calculated lift distribution, and to
determine the lift distribution for a given wing. Further, it became possible
to predict the functional relation of lift to the angle of attack and later on
to explain the dangerous phenomenon of stalling in flight. Investigations
into the stability of the boundary layer made it possible to delay stalling in
flight through the design of more appropriate wings and the influence of
the boundary layer by means of blasts of air, suction and artificial
turbulence.

This, then, is a theory whose propositions always have a direct technical relevance and which is also open to further development in line with the state of technology. The use of biplanes and propellers led to the elaboration of the multi-airfoil theory, in that higher air speeds first required a theory for swept-back wings, and then later a special airfoil theory which would take account of the problems of the sound barrier, and finally a theory for the field of supersonic flight.

These two examples present theoretical techniques which are both based on valid, fundamental physical theories, and are both guided by goals that are not already given within the framework of these fundamental theories (45).

A different kind of connection to the natural sciences can be found in more formal theories of technology, as for example in systems theory or in information theory (cybernetics). These utilize mathematics and logic to understand the formal structures of artificial information-processing systems. Originally developed as theories of technical systems, they were later applied also to natural structures.

At the outset information theory confronted the task of developing techniques for the transmission of messages, inventing various systems of coding, and transmitting messages without redundancy or distortion. Certain quantitative concepts such as information and redundancy were formed, and certain relations were established which made possible the theoretical solution of the optimisation problems. Information theory was later integrated into a general theory of information-processing systems. This theory now is applied not only in computer science but also in cybernetical models of biochemical, neurophysiological, ecological, physiological, and communicational structures. Technological progress has reached a stage at which its theoretical capacity yields special theories of natural processes as a sort of by-product.

The cases of fluid mechanics and chemical engineering have indicated how general scientific theories of nature such as hydrodynamics or classical mechanics became interconnected with attempts to solve technical problems by means of theoretical models such as the boundary layer or macrokinetics. In the case of information theory, the reverse also holds true: a general technological theory is here applied to natural phenomena. For example, the model of the double helix interconnects general information theory with genetics and the model of the metabolism of the cell interlinks information theory with the 'life' of the cell.

Both the development of sciences into special theories of technology and the development of technologies into special theories of natural structures show that after four hundred years of the development of science and technology their unification is no longer merely a philosophical project.

Notes

1. The history of the construction of the Cathedral of Milan, including all the construction and design problems, has been described in detail by Frankl (1945), Panofsky (1945) and Ackerman (1949).

2. The otherwise unknown mathematician Stornacolo gave the following solution: The width of the cathedral or base of the bilateral triangle is 96 brachia (54 m). This measure was already established with the beginning of the construction in 1386. This yields the height $h = 96/2 \times \sqrt{3} = 83.134\ldots$ Rounding it off to 84, Stornacolo could subdivide the cross-section into rectangular squares with lengths of 16×14 br. This subdivision made it possible to figure out all the relevant heights of the buildings by means of basic arithmetic, as is shown by the following diagram (from Ackerman, 1949, p. 89):

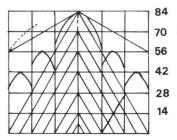

| | 84 |
| 70 |
| 56 |
| 42 |
| 28 |
| 14 |

Panofsky (1965, p. 63) explains the advantages of Stornacolo's solution:

> The problem was to reduce an irrational value containing the factor $\sqrt{3}$, to a reasonably accurate expression by means of a formula that had to fulfill three conditions. First it had to achieve its purpose by means of simple multiplication and division instead of by means of extraction of roots. Second, it had to be applicable without involving an extensive use of vulgar fractions of which the Middle Ages, not yet familiar with the decimal system, were understandably afraid. Third, it had to meet the requirements of architects who measured the sides of their triangles in large units ('unitates') and wished to have the results, viz., the approximate values of the heights of these triangles, in small units ('quantitates'), one unitas, we remember, comprising 8 quantitates.

Looking at Stornaloco's formula from this point of view, we discover that it is very practical indeed. It operates like a slot machine: we throw in unitates, pull the lever and receive a reasonably accurate result in quantitates. For all we have to do is to multiply the given number of unitates (expressing the side of the triangle) by 100, multiply by 700, divide the result by 1010 – and we obtain a number which, when divided by 10, expresses with reasonable accuracy the height of the triangle in quantitates. Calling the number of unitates *n*, we can rewrite the formula in general terms:

$$10h \; \frac{n100 \cdot 700}{1010}.$$

If *n* is 6, that is to say if the side of the triangle is 6 unitates or 48 quantitates, 10*h* is, as we have already seen, 415 (85/100), and it is at once evident that *h* itself is a little more than 41½ or, as Stornaloco would say, a little less than 42. If *n* is 12, that is to say if the side of the triangle is 12 unitates or 96 quantitates,

$$10h \; \frac{12 \cdot 100 \cdot 700}{1010} = 831 \; \frac{69}{101}$$

and it is equally evident that, in this case, *h* itself is a little more than 83, or, to speak with Stornaloco, a little less than 84. If *n* is 4, the result for 10*h* is 277 (27/101), and for *h* itself a little more than 27½ or a little less than 28; and so forth. We now perceive the advantage of writing out the formula 10*h* instead of for *h*. For in doing so, we automatically obtain what would be called today the first decimal point of the final value, the 0.5 after 41, the 0.1 after 83, the 0.7 after 27, etc.; so that the fractional rest can be neglected for all practical purposes.

3. Ackerman (1949) has reprinted the respective parts of the records of the Milan architectural workshop. See p. 109.

4. This phrase is Ackerman's, p. 104. The explanation of the Milanese success is to be found in the fact that Medieval statics overdesigned the reinforcements of the construction, and, in fact, worked more from the criteria of geometry than from those of statics.

5. The conflict between the Gothic and the Romanesque tradition took place in other towns as well. A short documentation of the disputes at Florence in the second half of the Fourteenth century is given by Prager and Scaglia (1970), p. 2ff.

6. See Gille (1966), p. 179ff.

7. For example For. III 43 v. and C.23 v. (Ed. Zamboni, p. 33). For further interpretation see Krohn (1977), p. 95f.

8. For a more detailed description, see Olschki 1927, p. 76f. Tartaglia uses another significant phrase, the 'practical speculation'. See for example the title of the ninth book of his *quesiti et inventioni* (1554) and the title of the sixth part of the *trattato de numeri et misurie* (1560).

9. For Paracelsus see Pagel (1958); for Paré see Zilsel (1945); for Vesalius see Farrington (1938).

10. For Gilbert see Zilsel (1941).

11. See Averdunk and Müller-Reinhard (1914).

12. Descartes (1967), p. 119.

13. ibid., p. 121.

14. For the compounding of the idea of the law of nature with technical rules by Descartes, Bacon, Newton, Hooke and others see Zilsel (1942), p. 86.
15. Leonardo, C. A. 119 v.a. (Ed. Zamboni, p. 13). Our italics.
16. Leonardo, C. A. 337 r. 6 (Ed. Zamboni, p. 17). Our italics.
17. Leonardo, G. 47 r, [Ed. Zamboni, p. 27]. One might consider whether the above quotation refers to the *social* purpose of discovery. It is more probable, however, that Leonardo's position was to seek the *causa finalis* more in artificial things than in things natural, as indicated by the following remark: "You investigators therefore should not trust yourselves to the authors who by employing only their imagination have wished to make themselves interpreters between nature and man, but only [to the guidance] of those who have exercised their intellects not with the signs of nature but with the results of their experiments". 1 102 r [Ed. Zamboni, p. 29]. All translations into English of Leonardo's notes as cited in this paper are taken from Leonardo, Ed. by MacCurdy (1938).
18. Cassirer (1927) p. 161ff.
19. Piaget (1975), p. 82. For a more fully elaborated interpretation of the Scientific revolution of the Renaissance in the light of Piaget's theory of cognitive development, see Krohn, (1977).
20. Kant, *Critique of pure reason*, B XIV.
21. For an analysis of the social structures that underlay the founding of the Royal Academy see van den Daele (1977).
22. See Heckscher (1932), Vol. I, 165ff.
23. See Habermas (1968). Habermas overstates the significance of the "functional circuit of instrumental-rational action" for early modern science. The sweeping analogy between instrumental action and modern scientific ideals of explanation does not reach the concrete level of theory formation.
24. See Boas-Hall (1974).
25. See Mason (1961), p. 311; Clark (1970).
26. The fact that many and particularly the more general theories of modern science are theories about the functions of instruments leads to metatheoretical consequences which are in P. Janich's contribution to this volume 'Physics – Natural Science or Technology?', p. 3ff.
27. It is very important to realize that the development of this technology particularly useful for scientific purposes was independent of the technology employed in the productive forces of the economy, though, of course, there existed a small market for scientific instruments.
28. The precision of the pendulum clock was useless for solving the most urgent problem of time keeping of that period: time keeping at sea.
29. See Koyré (1948).
30. See Middleton (1966). p. 28ff. and 48ff.
31. In the making of thermometers it was first proposed that all scientists should procure their thermometers from one and the same instrument maker. The problem of universality was later resolved fairly simply by establishing two fixed points. But even this solution presupposes the use of pure mercury and the regularity of the capillary tube.
32. Among the best examples of Descartes' deductive approach to experience is the *Traité*

de L'homme. Here even the interaction between the determination of the will by the soul and the activities of the body is mechanistically constructed.

33. See Dijksterhuis (1956) p. 534 and 547f. On the other hand, Newton was not able to work out his mechanistic program for all levels of nature. Material qualities and chemical processes were not covered by the "principia mathematica" and the general principle of gravitation was not in line with mechanistic technology, which excluded *actio in distans.*

34. See Aitkin, *Science, Technology and Economics: The Invention of Radio as a Case Study.* In this volume, p. 89 ff.

35. The role of science in the industrial revolution is emphasized, for instance, by Schofield (1963) and Musson and Robinson (1969).

36. The non-scientific resources of invention, of course, remain extremely important in the nineteenth century. There are examples showing that the problems of scaling-up, of rendering procedures more precise, and of controlling disturbing effects, are resolved more or less satisfactorily by the empirical method, occasionally independent of already existing offers from theoretical science. Thus Stephenson optimized the use of steam power in the steam engine by practical inventions, and without any recourse to Carnot's theory of the motive power of heat. Similarly, in 1856 Bessemer patented a new principle of steel production which he had discovered through systematic experiments based on a theoretical explanation. But at the same time the same principle was discovered by the American Kelly in empirical-practical experiments. A parallel example is provided by Layton's essay in this volume on 'Millwrights and Engineers', p. 61 ff. But for the production of synthetic dyes, by the end of the nineteenth century the techniques and concepts of organic chemistry were an indispensable condition.

37. See J. Bernal, 1963. 107 pp.

38. The problem of the steadiness of locomotives at high speed was highly relevant to save Stephenson's main-line long-boiler engine of 1841, especially after a long-boiler engaged in a demonstration before the Gauge Commissioners in 1846 left the road and turned over (Snell 1971, p. 100).

39. Obviously there was a demand for science to solve technical problems even before the nineteenth century. Thus, for instance, when in pre-revolutionary France the production of saltpeter, which was the major constituent of gunpowder, had diminished, a prize contest with a view to an improvement of saltpeter production was offered under the auspices of the French Academy. For Lavoisier, who had formulated the conditions for the prize competition, it was clear that the replacement of the existing procedure – the collection of saltpeter was practiced generally by scraping walls and sweeping stables – was primarily a scientific problem. "The chemists have not established an entirely satisfactory theory of the principles of nitric acid, on its true origin and the mode of its formation" (Multhauf 1971, 166). But the program did not bring a solution, as the analytic capacities of eighteenth-century chemistry were not equal to the task. Thomas, on the other hand, could by 1870 proceed on the basis of the much more sophisticated techniques of elementary analysis and of the theory of acids and bases.

40. Pasteur, *Oeuvres* V, 215.

41. The contingency and broad variety of the relations between science and technology is amply demonstrated by a series of case studies in this field. See for example Mayr(1975)

on the invention of the high speed steam engine, and Mayr's essay (1971) on the development of the technology of speed regulation. See also Layton's contribution to this volume (pp. 61 ff.). More classificatory and comparative ideas are developed in Layton (1971) and in Layton's and Mayr's essays in *Technology and Culture*, Vol. 17, Nov.. Oct., 1976. This volume is dedicated to the historical analysis of science and technology and is focused on the second half of the nineteenth century.

42. The presentation of chemical engineering is based on a study by Buchholz (1974).
43. Eucken and Kirschbaum, both pioneers of the theoretization of chemical engineering, pointed out in 1934 that "in the construction of apparatus a vast and fruitful research field lies barren in that it is possible to clarify the outstanding question through systematic research". Cited after Buchholz (1974), 10.
44. The presentation of fluid mechanics draws on a case study by Böhme. It is to be published in Böhme *et al.* (1978).
45. The application of scientific theories to goals that are external to the object area explicitly defined by these theories demands and leads to the formulation of new theories, which function, so to speak, as bridges between general theories and new goals. Under the title 'finalized sciences' we have analyzed the structure of these theories in Böhme *et al.* (1976). Case studies related to the development of theories whose development is guided by an external goal orientation are to be published in Böhme *et al.* (1977a). Consequences of the political control of scientific developments are analyzed in van den Daele *et al.* (1977).

Bibliography

Ackerman, J. S., 'Ars sine scientia nihil est'. Gothic Theory of Architecture at the Cathedral of Milan, in *The Art Bulletin* **XXXI** (1), March 1949, p. 84f.

Artz, F. B., *The Development of Technical Education in France 1500–1850*, London: MIT Press, 1966.

Averdunk, H. and Müller-Reinhard, J., *Gerhard Mercator und die Geographen unter seinen Nachkommen. Petermanns Mitteilungen*, Ergänzungsheft, Nr. 182, 1914.

Bernal, J. D., *Science and Industry in the Nineteenth Century*, Bloomington: Indiana University Press, 1963.

Bernal, J. D., *Wissenschaft (Science in History)*. Reinbeck: Rowohlt, 1970.

Boas-Hall, M., *Science in the Early Royal Society*. Paper presented to the Summer Meeting of the British Society for the History of Science, Leeds, 2–4 July 1974.

Böhme, G., Die kognitive Ausdifferenzierung der Naturwissenschaft, Newtons mathematische Naturphilosophie, in Böhme *et al.*, 1977, p. 237.

Böhme, G., Van den Daele, W., Hohlfeldt, R., Krohn, W. and Spengler, T., *Die gesellschaftliche Orientierung des wissenschaftlichen Fortschritts*. Frankfurt: Suhrkamp, 1978 (forthcoming).

Böhme, G., Van den Daele, W. and Krohn, W., 'Alternativen in der Wissenschaft', *Zeitschrift fur Soziologie* **1**, 1972, pp. 302–316.

Böhme, G., Van den Daele, W. and Krohn, W., 'Finalization in Science', *Social Science Information* **15**, 1976, pp. 307–330.

Böhme, G., Van den Daele, W., and Krohn, W., *Experimentelle Philosophie*. Frankfurt: Suhrkamp, 1977.

Buchholz, K., *Zu Stand und Entwicklung der Verfahrenstechnik*, Ms., Frankfurt, 1974.

Cassirer, E., *Individuum und Kosmos in der Philosophie der Renaissance*, Leipzig and Berlin, 1927. New printing: Darmstadt: Wissenschaftliche Buchgesellschaft, 1974.

Clark, G., *Science and Social Welfare in the Age of Newton*, Oxford: Clarendon Press, 1970 (1937).

Van den Daele, W., 'The Social Construction of Science', in Mendelsohn, Weingart, Whitley (eds.), *The Social Production of Scientific Knowledge*, Dordrecht: Reidel, 1977.

Van den Daele, W., Krohn, W. and Weingart, P., 'Political Direction of Scientific Development' in E. Mendelsohn, P. Weingart, and R. D. Whitley, (eds.), *The Social Production of Scientific Knowledge*, Reidel, Dordrecht, 1977.

Deffossez, L., Les savants du XVIIe siecle et la mesure du temps', Lausanne: *Ed. du Journal Swisse d'Horlogerie et de Bijouterie*, 1946.

Descartes, R., *The Philosophical Works* (translated by E. S. Haldaue and G. R. T. Ross). Vol. I, Cambridge, Mass.: U. P., 1967 (1911).

Dijksterhuis, E. J., *Die Mechanisierung des Weltbildes*, Berlin, Heidelberg: Springer, 1956.

Farrington, B., 'Vesalio and the Rain of Ancient Medicine', In: *Modern Quarterly*, Vol. I, 1938, p. 23 ff.

Frankl, P., 'The Secret of the Medieval Masons', in *The Art Bulletin*, XXVII (1), March 1945, 46f.

Gille, B., *Engineers of the Renaissance*, Cambridge, Mass.; MIT Press, 1966.

Habermas, J., *Erkenntnis und Interesse*, Frankfurt: Suhrkamp, 1968.

Hall, A. R., 'Merton Revisited or Science and Society in the Seventeenth Century', in *History of Science*, Vol. II, 1963.

Heckscher, E., *Der Merkantilismus*, 2 vol. (translated from Swedish by G. Mackenroth) Jena: Fischer, 1932.

Krohn, W., 'Die "Neue Wissenschaft" der Renaissance', in Böhme *et al.*, *1977*, p. 13ff.

Koyré, A., 'Du monde de la peu pres à l'univers de la précision', in *Critique* 4 (28), pp. 809–882.

Layton, E., 'Mirror-Image Twins: The Communities of Science and Technology in 19th Century America', in *Technology and Culture* 12, 1971, 562ff.

Leonardo da Vinci, *Philosophische Tagebücher*, Zusammengestellt und übersetzt von Giuseppe Zamboni, Hamburg: Rowohlt, 1955.

Leonardo da Vinci, *The Notebooks*. Engl. (translation by E. MacCurdy), New York: Reynal Hitchcock, 1938.

Mason, S. F., *A History of the Sciences*. Rev. ed., London: Macmillan 1971.

Mayr, O., 'Victorian Physicist and Speed Regulation: An Encounter between Science and Technology', in *Notes and Records of the Royal Society of London* 26 (2), 1971, p. 205ff.

Mayr, O., 'Yankee Practise and Engineering Theory: Charles T. Porter and the Dynamics of the High-Speed Steam Engine, in *Technology and Culture* 16 (4), 1975, p. 570ff.

Middleton, W. E. K., *A History of the Thermometer*, Baltimore, Maryland : Johns Hopkins, 1966.

Mittelstrass, J., *Neuzeit und Aufklärung*, Berlin, New York: De Gruyter, 1970.

Multhauf, R., 'The French Crash Program for Saltpeter Production, 1776–1794', in *Technology and Culture* 12, 1971, pp. 163–181.

Musson, A. E. and Robinson, E., *Science and Technology in the Industrial Revolution*, Toronto: University of Toronto Press, 1969.

Olschki, L., *Geschichte der neusprachlichen wissenschaftlichen Literatur*. 3 vls., Leipzig and Halle 1919–1927. Reprint, Vaduz: Kraus 1965.

Pagel, W., *Paracelsus: An Introduction to Philosophical Medicine*, New York: S. Karger, 1958.

Panofsky, E., 'An Explanation of Stornaloco's Formula', in *The Art Bulletin*, **XXVII** (1), March 1945, 61f.

Pasteur, L., *Oeuvres completes*, Paris: Masson, 1922–1939.

Piaget, J., *Die Entwicklung des Erkennens II. Das physikalische Denken*, Stuttgart: Klett, 1975.

Prager, F. and Scaglia, G., *Brunelleschi: Studies of his Technology and Inventions*, Cambridge, Mass.: MIT Press, 1970.

Redtenbacher, F., *Die Gesetze des Lokomotivenbaus*, Mannheim: Bassermann, 1855.

Scheler, M., *Die Wissensformen und die Gesellschaft*, Bern, München: Franke, 2. Aufl., 1960.

Schofield, R. S., *The Lunar Society of Birmingham. A Social History of Provincial Science in Eighteenth-Century England*, Oxford: Clarenden Press, 1963.

Smith, C. S., *A Historical View of One Area of Applied Science – Metallurgy*, in National Academy of the Sciences, *Applied Science and Technological Progress*, 1967.

Snell, J. B., *Mechanical Engineering: Railway*, London: Longman, 1971.

Weizsacker, C. F. von, *Die Einheit der Natur*, München: Hanser, 1971, p. 193ff.

Zilsel, E., 'The Origins of W. Gilbert's Scientific Method', in *Journal of the History of Ideas*, **II**, 1941, S. 1 ff.

Zilsel, E., 'The Genesis of the Concept of Physical Law', in *Philosophical Review*, **LI**, 1942, S. 345ff.

Zilsel, E., 'The Genesis of the concept of Scientific Progress', in *Journal of the History of Ideas*, **4**, 1945, S., p. 325ff.

THE RELATION BETWEEN SCIENCE AND TECHNOLOGY — A SOCIOLOGICAL EXPLANATION

PETER WEINGART

Universität Brielefeld

1. The Methodological Relevance of a Comparative Analysis of Science and Technology

It is surprising that so far little interest has been paid by the sociology of science to the study of technology, let alone to the comparative analysis of science and technology and the relation between these two areas of knowledge production (1). The sociology of science is concerned with science, and this term more often than not has the meaning of pure academic science. Even though the range of analysis has widened in recent years the sociological fascination is focused on the phenomenon of science as an institutionalized subsystem of society, differentiated from similar systems such as technology and hardly touching upon the classic professions such as medicine or law. This is surprising if one considers some of the major theoretical concerns of the sociology of science, namely to determine the respective import of 'internal' and 'external' factors in the development of scientific knowledge, or the specific *differentiating criteria* of *systematic* and *everyday knowledge* or the particular *conditions* of the production of scientific knowledge.

Part of the explanation for the sociologists' reluctance to include technology as an area of analysis parallel to that of science lies in their very conception of these terms. As the best available case in point one may take the writings of Robert K. Merton. In his early study on *Science, Technology and Society in Seventeenth Century England*, although differentiating between science and technology as between theoretical and empirical practical knowledge, he analyzed various inter-dependencies between them. In his subsequent work he shifted the focus to the

Krohn/Layton/Weingart (eds.), The Dynamics of Science and Technology.
Sociology of the Sciences, Volume II, 1978. 251–286. All Rights Reserved.
Copyright © 1978 by D. Reidel Publishing Company, Dordrecht, Holland.

functional prerequisites of 'science', thereby not only losing sight of the institutional and motivational interplay of practical and theoretical determinants of subject selection, but at the same time relegating technology to the realm of the purely practical, extra-scientific (2).

Another characteristic of the prevailing conceptualization of science and technology is, of course, the differentiation between scientific discovery and technological invention with which not only different types of activity (in the case of the latter the "new combination from the 'prior art', i.e., ideas previously known . . .") are associated but also different intellectual objects, i.e., types of knowledge. Inventions are taken to be triggered either by scientific discoveries or by non-scientific factors (3).

These and similar conceptualizations are legend in sociological literature (and in the history of science). With respect to the motivational forces the differentiating criterion for technological pursuits is the lack of disinterestedness, in contrast to science which is defined by this norm. Intellectually, science, i.e., pure, fundamental science is reserved the privilege of producing new knowledge which is based on the assumption, or rather definition, that only the discovery of universal laws of nature demarcates progress in knowledge (4). This leads to the conclusion, to cite Price as an outspoken protagonist of this line of thought, that the relation of science and technology *cannot* change as the research frontier is represented by concrete individuals and the transfer of knowledge into technology always takes at least one generation of training due to the lag of communication (5).

As there seems to be such a solid consensus over the disparate nature of science and technology in the sociology of science, what reasons are there to contest the principal tenets of this consensus? Scepticism toward the sociological conceptualization of technology and its relation to science is induced primarily by historical analyses and by studies that focus on epistemological aspects of this relation. Merton's message and thrust of analysis in his famous dissertation of four decades ago, namely to regard science as being in reciprocal relation with its environing social and cultural structure has been taken up by economic historians, among others, to whom the relation of science and technology is of relevance for the determination of the factors of economic growth, but who thereby provide important illustrative material for the theoretical concerns of the sociologist of science. Although, as Peter Mathias writes, the debate over the role of science in technical change, over the relation of the Scientific and the Industrial Revolutions is, in the last analysis, unresolvable because "scientific proof of causation is impossible", it is the historians who remind the sociologists that "much of the debate . . . hinges on what interpretations are given to the word 'science' and 'empiricism', or to 'pure' and 'applied' science" (6). It is in the context of economic theory that both historical and systematic analyses of modern configurations of the relation of science to technology have been initiated which have brought about a number of important insights challenging the evidently oversimplified dualistic views.

These historical studies draw attention to a methodological principle almost forgotten in sociology, namely that concepts are subject to historical change themselves just as are the phenomena they are meant to describe. Systematic analyses such as the attempts to trace and quantify the respective inputs of science and technology into selected 'events of innovation' reveal the crucial role of defining demarcation criteria and the difficulty of doing so perhaps more because of their methodological shortcomings rather than their results. They almost inevitably lead to a careful consideration of epistemological connections between science and technology. Finally, both types of analysis point to the fact that this relation may have to be regarded as changing in the course of history, both in its institutional and its intellectual configurations.

The account of the relation between science and technology that these analyses suggest is therefore more differentiated than the simplistic ones that dogmatic sociologists and historians want us to accept. It seems that among those undertaking the analyses just referred to a consensus has emerged about the evolution of science and technology which may roughly be characterized as follows: during the *first phase*, i.e., the Scientific Revolution neither technology nor science are institutionalized as separate social systems. Modern science, which is just emerging as a specific method of coping with reality, chooses many practical problems as subject matters thus inevitably following the footpaths of technology.

A *second phase* is characterized by the institutionalization of science eventually leading to the differentiation of science and technology. In cognitive terms this is the phase when science achieves its autonomy meaning that its subject matters are largely delineated according to theoretical considerations.

A *third phase*, finally, can be described as the reversal of the second. Science and technology experience a rapprochement, only now it is science that generates new technologies (science-based technologies) or, in other words, technological innovation is generated by the application of scientific method to technical problems. This implies at the same time that science has reached a stage where it can be oriented to practical goals (7). The characterization of this phase as that of the 'scientification' of technology (as well as of other areas of society) is perhaps the one about which there is most debate.

Behind this descriptive account there is what may be termed the differentiation-scientification thesis. Its methodological foundation is not entirely satisfying either, primarily because it is based on singular examples whose selection can be disputed. Also, some of the methodological shortcomings lead to contradicting results (8). One major problem is that of periodization, the other that of demarcating unambiguously the units of analysis. Because of all these difficulties, to cite Peter Mathias again, "all participants in the debate can rest happy in the awareness that their contributions, whether of more data, new hypotheses or judicious trimming between old ones, cannot be absolutely discounted as provably false or demonstrably irrelevant" (9).

While this affords us the necessary comfort for speculation, it must not conceal the fact that the historical debate lacks a theoretical scheme that serves to structure the plethora of historical data. (This does not mean that the historical analysis is entirely atheoretical, however the theoretical criteria that guide the selection of data and the conceptualization of 'facts' mostly remain implicit.) We want to develop such a theoretical scheme so that, if it is convincing, it may help to provide a coherent account of an otherwise more or less arbitrary selection and interpretation of data. The goal here is to develop a sociological explanation of the 'differentiation-scientification'-thesis, not – especially in view of the competence and space available – another historical analysis. Of course, such an explanation will also be subject to challenge by historical data that are overlooked and/or cannot be explained within its framework. But if it structures the ensuing debate it will have achieved much of its purpose.

We take the 'differentiation-scientification' thesis as a point of departure for several reasons. First because, as we have pointed out, it is the result of a more sophisticated methodological approach to the problems of conceptualization and interpretation of historical data than are otherwise available. Second, although at first the thesis seems to contradict some basic tenets of sociological theory, it can be shown that by taking a sociology of science approach these contradictions can be dissolved. Third, the thesis is of great interest both to the analysis of social change and to the problem of the relative weight of intellectual and practical determinants in the constitution of scientific concepts, instruments and subject matters. For the social study of science and of knowledge in general, the interaction of science and technology is of utmost importance. Both science and technology are knowledge systems that are focused on the discovery of or at least involve the laws of nature. They are different with respect to the relevance criteria that determine the choice and formulation of problems, the delineation of subject matters and the strategies of research. Thus, it becomes evident that a comparative analysis of science and technology and their interrelation will provide important information about the operation of such criteria of relevance in processes of knowledge production.

Before we embark on the presentation of the explanation proper, some assumptions that guide it have to be made explicit. We do not presuppose distinct concepts of science and technology but, instead, start from the notion that any social system needs to interact with nature and find ways to use or dominate it just as much as it has to regulate its social relations, this presupposes and generates systems of knowledge, in the simplest form experience. Such systems of knowledge are, analytically speaking, identical with respect to their functions of orienting and ordering action whether they be religious myths, cultural values, experience-based simple technologies or theoretical science. Each of these systems of knowledge interpret the relation of man or society and nature in different ways. In reality they are not mutually exclusive and do in fact exist side by side, but they may be competing and one may prove to displace or dominate the other. To explain this process by the superiority of one system of knowledge over another involves the difficulties of teleological models of evolution. To account for the prevalence of one over another at any given moment of time may be done by resorting to the notion of 'relevance criteria' which are institutionalized and which regulate the production and select the 'appearance' of knowledge (10).

These relevance criteria are analytically speaking the 'common denominator' which allows a comparison of different systems of knowledge, between different cultural or social contexts, and within identical contexts over time. The idea is that these criteria exist in any society, such as 'truth', 'well-being', 'security of life', etc. but that they are interpreted differently depending on the available knowledge, resources and cultural development. Also, they may assume differential weight with respect to their impact on social processes. They are the analogue to social institutions in that they structure cognitive orientations, more concretely processes of research and learning by constituting rules of problem selection and of determining problem solutions. By employing this concept 'scientification' will be interpreted to be the gradual gain in supremacy of the criterion of scientific truth which does not replace others but changes their interpretation. Economic wellbeing or health are no longer relevance criteria organizing the production and selection of knowledge as such but only in conjunction with scientific truth, i.e., only when it can be established that the pertinent knowledge can be shown to be true in the scientific sense. (For that reason non-scientific types of knowledge are discredited even though they may be 'successful'.)

In a full fledged theory of social knowledge in which all knowledge systems are treated as analytically equal, the relevance criteria, their relative impact and change over time would be seen as the explanatory variables accounting for the translation of social processes into the development of knowledge. Thus, an analysis of social, economic and political developments would play a major part. Such a broad approach hardly lends itself to produce more than suggestive results, however.

The crucial problem is to explain the dynamics of the differential development of science and technology, the sequence of differentiation and scientification as it is roughly described with the three phases mentioned above. The basic assumptions underlying the explanation of this process are the following: Having said that in any society the need to interact with nature generates systems of knowledge a further assumption is that any society organizes the production and the diffusion of knowledge both intra- and intergenerationally. Secondly, changes in any of these dimensions do not necessarily lead to simultaneous changes in the respective others, albeit they become limiting factors which account for an overall pattern of development – that of differentiation and scientification. Thirdly, the entire process, i.e., the different patterns that the system of production and diffusion of knowledge assumes, may be taken to be a changing configuration, historical interpretation and emphasis of relevance criteria.

The explanation that we intend to develop has to remain highly simplified. It will encompass several levels of analysis. It will be argued that in cognitive terms the introduction of modern science, i.e., the 'new method' of systematizing empirical knowledge first of all implies the process of differentiation of theoretical and practical concerns of knowledge production. Consequently this process leads to the penetration of the latter through the former, the scientification of 'non-scientific' or practical modes of knowledge production. On the institutional level it must be shown in which institutional framework the production and diffusion of knowledge takes place, how they are affected by cognitive change and how, in turn, different relevance criteria are mediated by them into cognitive processes. Here we proceed from the assumption that the development of the institutional framework does not follow the cognitive change simultaneously but that it does reflect the differentiation-scientification pattern as well. First, the functions of knowledge production for practical purposes are differentiated from those of the production of knowledge *per se*. Following that, the mediation of the differentiated contexts of knowledge production becomes institutionalized. The cognitive and institutional processes seen in conjunction reveal 'scientification' as the gradual displacement of every-day experience through systematic knowledge. This process results in the restructuring of the social patterns of problem perception and learning which is to be interpreted as a shift in the differentiated impact of different relevance criteria.

2. Pre-paradigmatic Science and Craft Technology

When considering the emergence of modern science in the Renaissance and its relation to technology, it must be borne in mind that up to that time a notion of science and technology or handicraft already existed and their interrelation as well as their respective social valuation had undergone various changes during previous periods. At the end of the fifteenth century, however, cognitive changes occurred that, temporarily at least, disrupted the link between the established systems of knowledge and the social structure.

At this time the ideal of the crafts of step by step improvements of the state of the art was linked up with the humanist ideal of individual fame and the achievement of ultimate perfection. The notion of a 'common good' that became an orienting value for the arts and sciences pointed toward the dissolution of the guilds which was based on the secret training of experience. The idea of progress implied *continuous search* ('Weitersuchen': Dürer) beyond everyday experience and thus the perfection of inventions which are at first always insufficient. This concept of a search for insights based on experience which nevertheless allows the perfection of the knowledge already available was extended to knowledge in general and the traditional institutional differentiation of theoretical and technical knowledge was abandoned (11).

The emergence of the 'new science' was no sudden event. The professional group of artist-engineers such as Brunelleschi, Ghiberti, Leonardo and Dürer who worked as painters, sculptors, architects and engineers was characterized by their distinct empirical and experimental procedure. But it took from the fifteenth to the seventeenth century for these methods to become more generally accepted (Gilbert, Galilei, Francis Bacon) (12).

There existed "rival images of knowledge" such as Humanism and Syncretism, and although they shared the experimentalist attitude "experience of the ancients as quoted in the sources was deemed equivalent, nay more reliable, than direct experience" (13).

The 'new science' and what were its legitimate subject matters cannot be isolated from the religious convictions of the time. One of the major developments was the separation of body and mind which served to neutralize the religious resistances against the experimental analysis of bodies of any kind (14). A very lucid expression of the delineation of legitimate areas of inquiry and of the relevance criteria involved is to be found in Sprat's passage. Thus, of the possible "objects of men's thoughts", God, Men or Nature, the natural philosophers are concerned with divine things only insofar "as the Power, and Wisdom, and Goodness of the Creator is display'd in the admirable order, and workman-ship of the Creatures". The Subjects of God and the Soul are forborne but "in all the rest they (the natural philosophers – P.W.) wander, at their pleasure: In the frame of *Mens bodies*, the ways for strong, healthful and long life: In the *Arts of Mens Hand*, those that either *necessity, convenience,* or *delight* have produc'd: In the *works* of *Nature*, their helps, their varieties, redundancies, and defects: and in bringing all these to the *uses* of *humane society*" (15).

If one follows Rossi the unity of truth and utility which is apparent here was also the core of Bacon's programmatic, so that Bacon's philosophy thus "took a deliberately strong position against the idea of a separation and opposition between technics and science, manual and intellectual work and mechanical and liberal arts" (16). The precondition of this unity is the adaption of the 'new method', i.e., the empirical and systematic study of phenomena. Only then do the twin human intentions of knowledge and power coincide in a single intention because "that which is considered a cause in the theoretical sphere is considered a rule in the operational sphere" (17).

The same idea is expressed in a similar way by Descartes when he writes that the practical philosophy leads to useful knowledge about the forces of fire and water, about the air, the stars and all other bodies that surround us in just the same way as we know the techniques of the craftsmen so that we can use them for all goals for which they are fit (18). This concept, where nature can be transformed into something that does not exist but that is possible, where the goal of the discovery of nature is not an end in itself, but the construction of facts according to rules which stake out the realm of the possible, comprises science and technology. The knowledge about nature has become methodologically and factually identical with the experimental and deductive construction of nature; science and technology are therefore identical (19).

These few references to conceptions of science and technology in this epoch may suffice to make some points about the cognitive situation at the time. The emergence of the 'new science' is, above all, the emergence of a new method to organize and systematize experience and thus knowledge. Thus, practically all traditional knowledge both of nature and that which had been accumulated in the arts and crafts became the object of study by means of the new method. This accounts for the fact that so many technical problems, the pumping of water, the determination of longitude, the ballistic curves of projectiles, the optimal shape of ships' hulls etc. were incorporated into the concern of science. The methods used as well as the objects studied were often not very far removed from daily experience and could not only be understood by specially trained scientists but by a great number of people with relatively little technical training (20). At the early stage of this development science, first of all the physical sciences, proceeds in a descriptive, often taxonomic fashion, even though in specific areas there are very early examples of theory building, e.g., the mathematical deduction of theories from phenomena such as Newton's law of gravitation. Science was 'pre-paradigmatic' in Kuhn's term. However, while the 'new method', the Bacon-Descartes quest for certain knowledge, had forced a temporary merger of science and technology, or to put it more correctly, an identity of subject matters, the very method implied the search for universal principles. According to Elkana one of the objects of Newton's revolutionary methodology was that the "transference of the certainty of facts to the 'principles' deduced from them conferred upon these principles not only the former's certainty but their logical status as well" (21). From the very beginning the 'new science' thus contained orienting patterns that would drive its development beyond the realm of immediate experience, i.e., that which could be obtained within the context of everyday life. But for science to become virtually a competing mode of experience, already existing and firmly institutionalized systems of knowledge, above all religion, had to be replaced. A look at the early *institutional development* of science reveals how in this phase different social relevance criteria are still operative within a single institutional framework.

The central institution of the 'new science' is the academy and its functions and internal structure reflect the interrelation of science and technology at the time. With respect to the functions of the academy, the most important characteristic is the variety of purposes it serves. The 'Academia del Cimento', founded 1657 in Florence and one of the first of its kind, was the foundation of a sovereign (the brothers Medici). The decade of its activity was devoted to the development of, and work with, measuring instruments, such as the hygrometer, the pendulum, thermometers, experiments with the vacuum, studies of the effects of warmth and coldness on different objects etc. (22). In contrast to the 'Cimento' the Royal Society emerged from the spontaneous association of amateurs and scholars and if the work of the 'Cimento' may be considered primarily 'scientific' (taking into account the mentioned dual character of the objects of study) that of the Royal Society, in retrospect, dealt with both 'scientific' and practical problems (the latter being connected with commerce, mining and manufacture). The concern with practical problems primarily earned the society the Royal Charter even though it contained no provisions for financial support (23). While the Charter acknowledged that the members of the society convened in order "to confer about the hidden causes of things . . ." the statutes of 1663 reveal the Baconian unity of truth and utility, of theoretical and operational problems. Thus it was the business and design of the Society "to improve the knowledge of *natural things* and *all useful* Arts, Manufactures, Mechanic practices, Engynes and Inventions by Experiment . . . In order to the compiling of a complete system of solid philosophy for explication *all phenomena* produced by nature or art and recording a rational account of the *causes of things*" (24).

Sprat in his history of the society demarcated its functions against the old science as well as against the pursuit of immediate ends when he pointed to "a double prevention; both by endeavouring to strike out *new Arts*, as they go along; and also by still improving all to *new* experiments" (25). This is underscored by another characteristic that the Royal Society shared with the French Academy, i.e., the monopoly to grant patents. As early as 1662 the task of examining all applications for patents which obviously required the competency to understand scientific methods and principles as well as technical rules from its members was conferred on the society.

The diverse activities of the society, as they seem to us were carried out within one and the same organization and mostly by the same men. There existed no institutional separation between these activities and if Merton arrives at the result that between 40 and 70% of the research projects of the society (in the years 1661, 1662, 1686 and 1678) show *no direct* relation to practical needs this tells us more about the scope of the activities in modern terms than that it reflects the contemporary categorization of research areas (26). The criterion of institutionalization, therefore, was *not* a differentiation between theoretical and technical interests, between the search for truth and the solution of practical problems but the systematizing and structuring function of the empirical and experimental procedures which, at that time, could, in principle be applied to all natural and artificial objects.

The foundation of the French Académie des Sciences, in 1666, marked the compromise between a utilitarian and a more encompassing cultural orientation. The academy had the dual function of dealing with technological problems posed to it by the king and at the same time glorifying him through the support of the sciences (27). This dual function was reflected in the criteria of selection of new members which were clearly directed against the protagonists of rigid philosophical and religious convictions but also against amateurs superficially interested in the advancement of knowledge. The generous financial support from the state and the actual work in the academy point to a higher degree of institutionalization of what was then considered science. Under Louvois a more 'technical' or 'practical' orientation became effective but it entailed primarily the application of knowledge to the construction of gadgets for the king. The reorganization in 1699 then merely ratified changes that had occurred in the preceding years. The consulting function was codified and the academy received the monopoly to grant patents. Within the academy the tendency of viewing judgements on scientific and technological matters as one and the same academic activity prevailed (28). The membership criteria corresponded with this view in that only one who had made a name for himself either within the science he represented or through a discovery or the invention of a machine could be put up for election (Art. 13). Similarly the internal organization provided that each member was called upon to devote himself to a particular science but at the same time extend his studies to whatever was useful and interesting (Art. 22).

Perhaps the most striking exemplification of both the function of the academy and the explorative state of the knowledge it produced was the conviction that the key to progress was the rationalization of the crafts through the application of the true 'scientific method' which resulted in the publication of the 'Description des Arts et Métiers' originally initiated by

Colbert. Also, the 'Encyclopédie' helped to stimulate widespread enthusiasm for science and became an important element of the Enlightenment.

From this sketchy description of the academies, it is evident that in what may analytically be identified as the first phase of the development of modern science no institutional differentiation between science and technology (both terms taken in their modern meaning) existed. Also, for that very reason, it is inappropriate to speak of an institutionalization of science in the modern sense as the academics as the organizational locus of the 'new science' still comprised a multitude of competing (or rather not yet differentiated) criteria of social relevance. Truth and utility are considered to be identical or at least not mutually exclusive. The religious systems of thought determined the line between legitimate and illegitimate objects of inquiry and the value system on which the political authority of the sovereign rested did likewise.

This situation is illuminated by a look at the diffusion of knowledge. It must be remembered that the academy as an institution emerged as a new type of organization especially designed to serve the needs of the new experimental science and its protagonists, who, by definition were amateurs. The new institution was necessary because the traditional university was founded on different principles of production and diffusion of knowledge and consequently resisted the new concept of experimental science. (The situation was slightly different in Germany where university reforms succeeded very early – 1697 the new university of Halle was opened – thus retaining a strong institutional continuity which left the academies relatively unimportant as an institutional innovation as compared to their French and English counterparts.) Although there are marked differences in development, in structure and function between the French, British and German university the one common characteristic of this institution is that well into the eighteenth century it was primarily devoted to theological teaching on the one hand and classical literary studies on the other (29). The only two professions besides the clergy that were both institutionalized and had their counterparts in departments or chairs within the university were the legal and the medical profession. The newly emerging experimental sciences and arts were not incorporated into the university curricula until much later.

With respect to the occupational structure of society the new sciences and arts, as they were practised in the academies, did not have an impact at this stage (i.e., roughly until the middle of the eighteenth century) and also were not part of the existing learning processes. With the exception of the cases mentioned, no professions nor any formalized processes of learning existed through which the pertinent knowledge was transmitted. In terms of the social production and diffusion of knowledge the central mechanism was still that of traditional societies, namely that of experience being obtained in the pursuit of occupational and everyday activities. It is primarily this experience which structures learning processes. The second such mechanism was the generational change which as a structural characteristic of the transfer processes of knowledge corresponded to a stage of development where experience rather than systematic knowledge was transmitted. The following analysis will show that the structure of social learning processes underwent profound changes in the course of the development of science and technology. With this account of the beginnings of modern science, of the cognitive and institutional frame-work in which science and technology are not yet differentiated but where, with the 'new method' a new principle of the social organization of experience had emerged the stage is set for the analysis of the process of differentiation and scientification.

Concluding this chapter it is necessary to recall the caveat stated at the beginning, namely that the development described emerges from a specific constellation of traditions in which theoretical (philosophical) and practical concerns were differentiated. The institutional unity of science and technology was achieved although for many scholars the philosophical tradition seemed to be more important. The 'new method', however, became the common denominator even of diverging interests in the production of knowledge.

3. Paradigmatic Science and the Institutional Differentiation of Science and Technology

When describing the main characteristics of the second phase in the development of science and technology, it must be emphasized again that we are arguing on a level of abstraction which does not fit neatly into the concrete historical periodization of events. The phenomena and events that serve as examples partly overlap and partly occur in a long drawn-out sequence. The time span of what is considered here the phase of paradigmatization of science and institutional differentiation of science and technology reaches from about the latter half of the eighteenth to well into the twentieth century. The question is how is it to be explained sociologically that after a phase of the institutionally integrated pursuit of scientific and technical goals which were not conceptually differentiated these become institutionally separated.

Evidently it is impossible to trace comprehensively the social processes that no doubt must have at least contributed to the institutional separation of science and technology. Instead, we turn again to the cognitive dynamics of the development of science. This is not to be mistaken as an 'internalist' approach because, here, we are not interested in a sequence of specific ideas and their interrelations but in the 'scientific method' as an orienting principle that structures social perception and behavior and its consequences both for the development of knowledge and the social organization of its production. Rather than repeating here well known historical facts about the evolution of scientific ideas, an analytical point may suffice.

Characteristic of the new experimental science was the search for general principles, i.e., for rational accounts "of the causes of things". In terms of the processing of knowledge (or information) this is a procedural rule which constitutes an activity that is determined by the type of problems that are posed but not, *prima facie*, by the objects to which it is directed. Why, then, does this activity eventually lead to a separation of natural and technical objects and likewise to a separation of scientific and technical interests? The crucial point is that the search for causes transcends human artefacts and leads to the underlying principles of nature.

The substitution of empirical (i.e., trial and error) procedures by scientific rules was an enormous rationalization of technology. At the same time the search for causes and the pursuit of problem sequences continuously produced new knowledge about rules and principles.

As long as science and technology were not institutionally separated this indicated that causes for empirical procedures that as such were known, were searched. The differentiation is already latent when as a result of the solution of secondary problems, i.e., problems appearing in the course of the search but are not apparent at its outset, causes and principles are found that can be the foundation for new procedures. These 'secondary problems', constitute also an increasingly independent realm of the search for natural laws.

This does not mean that the search for natural laws is not determined by prior conceptions of them. In fact these conceptions, being empirical or theoretical interpretative frameworks that call "for further articulation and specification" emerge from the occupation with natural or artificial objects in the attempt to explain causes (30). It is after such orienting frameworks have become the basis of the consensus of a sufficiently large system of discourse that they can be said to determine the type of activity that Kuhn has termed 'normal science'. What Kuhn describes as the "foci for factual scientific investigation", the determination of facts that are particularly revealing of the nature of things "with more precision and in a larger variety of situations" and of those that "can be compared directly with prediction from the paradigm theory", and finally empirical work such as experiments to determine physical constants, to establish quantitative laws and to choose among alternative ways of applying the paradigm to a new area of interest further elucidate the particular nature of this activity (31). Once a paradigm has found acceptance, the process of its articulation, i.e., the search for general principles is set free, separated from problems other than those that are generated by the very activity itself.

This development of a distinct system of knowledge production takes a considerable amount of time. First the boudaries of the system can only be determined in cognitive terms, i.e., by the nature of the problems that are generated within it. Not until much later does this system become institutionalized. The cognitive separation is demonstrated by the development of instruments and analytical procedures. In many cases – at least in the early phases – these instruments existed already in the practical technical context such as the clock or the pump, for example. When they became the object of analysis in the context of the search for causes and general principles the context of their use, i.e., the purpose they served, was transcended and another set of relevance criteria structured the questions that were put about them.

In the case of analytical instruments and procedures this becomes evident in a shift of interest to such norms as precision, unambiguousness, universality and generalization (32). By following these norms the knowledge that is produced – both as a result of the analysis of instruments and their use for the analysis of natural phenomena – and the concept of experience of nature are fundamentally changed, from individual sensual observation on phenomena to experience mediated by rules and instruments (33).

Luhmann has pointed to an apparent paradox that is crucial for the understanding of this evolution. The new concept of experience implied by the experimental method restricts the conditions of truth, and this strategically placed scarcity of truth results in the immense increase of true information about the world. Abstraction, functional specification and differentiation of the medium of communication of truth are the basis for the success of modern science. The institutionalization of science as an autonomous system of communication, its differentiation from other criteria of relevance enables it to deal with high complexity without having to fear the consequences (34). This is, in effect, the essence of the differentiation of truth (and its derivatives) and utility as distinct criteria of relevance guiding the production of knowledge.

The conditions that are described as the paradigmatization of science, i.e., the emergence of a theoretical dynamics of evolution, may be said to be given first in mechanics or physics with Newton's *Principia*, then in chemistry after Lavoisier' discovery of oxygen and finally in physiology (or what became biology) with the abandoning of the concept of vital force. The time span that these developments covered is also that of the process of institutionalization of science and its separation from technology (as well as other systems of discourse). What, then, were the chief characteristics of the institutional processes?

The crucial criterion of the institutionalization of science and its differentiation from other social systems and their competing relevance criteria is the *role of research*, because it is through research that knowledge is produced and the contents of communication is determined. Although the academies of the seventeenth and eighteenth century were early forms of institutionalization of science, they remained feeble. Their quantitative capacity was limited. They comprised different functions, (however, lacking systematic teaching) and they failed to establish research as an autonomous professionalized activity. The new institutional framework which completed the differentiation between science and technology was the result of a peculiar constellation of political and intellectual conditions: the German university. At the time, it was only by the realization of Humboldt's idea of the university that the functional differentiation of science and technology was accomplished, while elsewhere, due to national differences, a closer connection continued to prevail. Thus, while the paradigmatization of science can be explained in terms of a cognitive dynamics the institutional configuration that it assumes is a historical windfall and by no means a necessity.

The ideas that were prominent in the debate over the reform of higher education in Germany around 1800 ranged from the established scholarship of the traditional university which was opposed to utilitarian reforms, to the enlightened intellectuals and politicians who expected an encyclopedic and utilitarian education from science. The more radical among the latter demanded specialized schools of higher education for different occupations according to the French model. The position in between represented various shades of the the idealist philosophy, being opposed to the enlightened utilitarian ideas, at the same time intending to give a new conception of science a new institutional framework (35). The neohumanist concept of education molded with the idealist concept of science, as formulated by Wilhelm von Humboldt, became the guiding principle of the university reform. In idealism science and education (*Bildung*) converge in a concept of education through science. The self realization of man is to be achieved only through "insight into the pure science". (The 'pure science', of course, was philosophy as critical reflexion.) With this idea of education, a reflexive concept of learning was introduced that provided the university with a unique function. The university is not the place where specific knowledge is transmitted – like in the schools – but rather the method of learning, or in Schleiermacher's words the learning of learning. University education is not organized as the specialized training for particular occupations but as a general education (*Allgemeinbildung*) which after having provided a general understanding of the world and society serves as the prerequisite for subsequent specialization. For the method of teaching this implied the presentation of the research process in which results were achieved rather than of the results themselves. For the demands on the university teacher it implied that he had to be a researcher as well. The teacher would obviously present his own research, the most advantageous method of presentation would obviously be the practical demonstration and active practice of research. With this the conception of the "unity of learning and research" is achieved (36).

The decisive consequence of these idealist principles institutionalized in the 'philosophical university' is the separation of science from its application in life. Science is no longer directly related to practice but serves society only *indirectly*. Knowledge and practice which Bacon conceived to be a unity now become *mediated*. Although the idealist 'Naturphilosophie' which in the subjection of the scientific disciplines under philosophy had become effective in the organization of the new university lost its influence after Hegel's death in 1831, research in the natural sciences continued to be conceived as 'pure' science. The institutional separation of science and its application as well as technology had become irreversible. Eminent scientists who had been critical of *Naturphilosophie* like Helmholtz and Liebig nevertheless adhered to the idea. The idea which had become institutionalized and shaped the organizational behaviour as well as the legitimizing ideologies was that research that is directed to the complete knowledge of natural laws had to follow their supposed inherent logic. Only in the second instance did that knowledge provide rules for technical realization. Technology was thus considered subsequent to science, it was its application (37).

The crucial innovation of the German university, and the crucial step towards the institutionalization of science and its differentiation from technology was the establishment of research at the university. This was important because it implied the exclusion of competing relevance criteria guiding the production of knowledge. With its particular dual function of producing knowledge and at the same time being the medium of learning, an autonomous universe of discourse was established. Research could be learned, it became an occupational role and constituted a profession which gained control over the evaluation of achievement, the criteria of access, and the recruitment and training of new recruits, all in the same institution. It is unique for this profession that it has no outside clientele which can serve as a reference group determining its actions and

evaluating them. The institutionalization of research at the university is
the institutionalization of 'normal science' (38). It is the crucial step in the
long process of the institutionalization of truth as a criterion of relevance.

At the time the German university was the only institution which
achieved the separation of theory and practice, or 'pure' science and
applied science and technology. France, for example, reveals a quite
different picture, and one where the separation was not achieved in the
same way. On the one hand, the system of specialized technical training
schools seems to have been the result of an early utilization of science by
the state and independent from the pre-enlightenment university. This is
evidenced by their being subject to the jurisdiction of different ministries
such as commerce (*École Centrale, École des Arts et Métiers, Écoles
Pratiques de Commerce et d'Industrie* etc.), public works (*École
Polytechnique, École des Ponts et Chaussées, École des Mines*), military
(*École de Génie d'Artillerie*) and agriculture (*Institut Agronomique, École
Forestière*). These schools, the foundation of which started in the second
half of the eighteenth century and culminated in the establishment of the
famous *École Polytechnique* clearly served the needs of the state in the
areas of military and civil engineering, and later were supplemented by
specialized schools for the needs of industry. However, they were not
limited to a purely technical training function but also fostered the natural
sciences. The *facultés des sciences*, on the other hand, although of a lower
standard than the *écoles* were primarily oriented to the teaching of the
natural sciences but established specialized institutes of applied science
designed especially to meet the local and regional needs of the utilization
of scientific knowledge. (The *facultés* were under the jurisdiction of the
ministry for public instruction, together with the *École Normale
Supérieure*.) Thus, both types of institutions comprise science and
technology or applied science. Their specialization is determined by
disciplinary and occupational differentiation, not by a supposed inherent
difference between pure and applied knowledge. Consequently, they did
not entail an autonomous research function, they were primarily devoted
to teaching while research still remained a matter of private initiative (39).

That the separation of science and technology as it was set forth with
the German concept of the university, was by no means immediately
realized even in that country becomes evident if one looks at the situation

of the specialized schools in Prussia. There, the *Artillerie- und Ingenieur-schule* educated military engineers, the 'Bauakademie' trained the technical civil servants. These institutions, too, were oriented more to the standards of science than to those of technology. The same is true for the Polytechnical Institutes in Prague (1806), Vienna (1818) and Karlsruhe (1825), due to the fact that they were primarily designed to educate technical personnel for the state. The first institution in Germany serving the needs of a supply of technically trained graduates linked to the beginning industrialization was the *Berliner Gewerbeinstitut* (1821) which was later merged with the older *Bauakademie* to form the *Technische Hochschule* in Berlin (1879) (40).

Although the early Polytechnical Institutes did not, in fact fulfill their goal of becoming the equivalent of the university for the "important class of the higher manufacturers, managers, and agents of commerce", they marked the beginning of a technical training which was both cognitively and institutionally separated from science. The differentiation of the technical institutions from the university was aimed at by pronouncing the 'technical principle' versus the 'principle of scholarship' i.e., the 'unity' or the 'essence of technology' was declared the organizing principle of these schools (41).

The subsequent development of the Technical Institutes in Germany is characterized by the debate over a (re-)integration with the university or the independent development of technical education and equal academic status with the university. This was reflected in intellectual efforts to amalgamate technology and science as well as attempts to introduce scientific methods into technology. One example is Redtenbacher's formulation of the 'laws of locomotive construction' of 1855 and his attempt to 'base the whole discipline of engineering on reliable rules" (*sichere Regeln*). Another example is Reulaux, who wanted to introduce the deductive method into engineering and aimed at a planned and directed scientific method of discovery (42).

The real emancipation of the technical disciplines occurred in the last third of the nineteenth century. It is characterized by the introduction of experimental teaching and research while the theoretical orientations which had led to a growing autonomy of the non-technical basic disciplines that had been established at the technical schools lost more and more of their influence. The introduction of the experimental technical laboratory at the *Technische Hochschule* in Munich (1868) provided the organizational prerequisite for the independence of this institution. With the laboratory the peculiarity of technology as a field of research was accepted, the "understanding of the diversity of practical conditions", which was only possible with particular methods, i.e., experiments on machines of full scale under conditions which correspond to a 'realistic operation' (43). When the *Technische Hochschulen* had finally achieved equal academic status to the universities at the turn of the century (by being granted the right to confer doctoral degrees), they did so by having developed a research conception of themselves, unique to technology. The mode of production of knowledge that was thus institutionalized was different, though, from that in the universities.

This development, which covered the entire nineteenth century, does not lend itself to an easy explanation. As Manegold has shown the institutional separation of scientific and technical training and research must be accounted for, at least in part, by the attempt of the engineers to achieve the same social status as the university, initially by emulating academic science, later by finding their own identity and giving it an independent institutional framework. This professional policy alone, however, cannot explain why, for example, the university did, in fact, serve as the frame of reference. In order to explain why a specific cognitive principle was formulated to justify the particular organizational prerequisites of teaching and research one has to take recourse to the dynamics of scientific and technical knowledge. National differences, again, point to the importance of preceding institutional arrangements as well as governmental policies which translate societal needs into structures of knowledge production and diffusion. Finally, the development of technical and applied research also has to be seen in light of the fact that from the middle of the century onward the demands on knowledge from industry became more explicit as industrial development experienced a surge. (It is indicative that during the phase when the theorization of technological research was propagated, industry began to establish its own research institutes.) Thus, it is probably inappropriate to attribute the institutional differentiation to any one factor. It is plausible to hypothesize, however, that the cognitive dynamics of systematic knowledge accounted for an increasing differentiation and specialization and that, at the same time, this tendency was mediated by different existing institutional arrangements or political conditions leading to different configurations of the interrelation of science and technology. In terms of our analytical framework it is the prevailing social, political and cultural context which determines how the competing relevance criteria affect the production of knowledge.

Concluding this analysis of the phase of paradigmatization and differentiation, we look at the change of the mechanism of the social production and diffusion of knowledge from a macrosociological perspective again. The difference between technical and academic learning was in contents only, not in principle. The experimental laboratory as a device for teaching is not to be mistaken as an indicator for a recourse of the engineering sciences to 'trial- and error'-procedures even though they may still play an important role. Engineering had become a subject that could be taught, i.e., both scientific and technical knowledge as it evolved through continuous, routinized research constituted an entirely new type of experience about the world. Likewise, in industry at the beginning of the nineteenth century the increasing importance of systematic-methodical approaches as opposed to the traditional empirical ones based on experience became apparent. It was the transition from 'empirical' to 'rational' management, from the master to the engineer. As the role of abstract, theoretical knowledge for the production process increased, a differentiation and formalization of an ever growing number of training functions became necessary that until then had been under the competency of the crafts. Because of its nature as universal, generalized and reproducible knowledge, it could be disconnected from the traditional mechanism of diffusion and transmission, namely learning by doing as it was institutionalized in the master-apprentice relation. In more general terms, then, experience became abstract, systematic and accessible through routinized and methodical research. This new type of knowledge called for, or made possible the formalization of learning. With the institutionalization of formalized learning processes a new biographical pattern emerges: a phase of learning is now sharply separated from the subsequent phase of application of knowledge. In this latter phase only informal learning processes take place. However, despite the biographical separation between learning and the application of knowledge the generation change as the structural characteristic of the transfer processes of knowledge remains intact. The institutionalization of formalized learning, then, is essentially a response to the previous introduction of the systematization of experience (the 'new science') and its subsequent institutionalization as autonomous research.

Seen in this context the differentiation of scientific and technical knowledge (research and training) appears once again as a transitional stage of evolution. Once the principle of systematic, universal and methodically (re-)producible experience was initiated, by virtue of its superiority over all other forms of knowledge production and its inherent cumulative dynamics, it had to develop into the major principle of the social organization of gaining and structuring reliable experience. Variations both in the temporal sequence and in the institutional framework in which it is embedded reflect the culturally specific resistance of other relevance structures. The differentiation of scientific and technical knowledge, both as occurring first in a specific national context and later on as evolving wherever academic research was institutionalized, may thus be interpreted as evidence of the *partial* realization of that principle. It follows from this that the differentiation may be expected to disappear as the principle of systematization of experience further overcomes the resisting structures.

4. The Scientification of Technology and Reflexivity of Social Practice

The third analytically distinguishable phase we characterized at the outset of this analysis is that of the 'scientification' of technology (and, in fact, of most other areas of practice). Such a description of the development of science and technology, with first indications of it becoming apparent around the second half of the last century can be convincingly supported by empirical evidence. On the surface, however, it seems to run counter to the general principle of social development, i.e., increasing differentiation. It will be shown that this apparent contradiction can be resolved.

At first, we look at the cognitive relation of science and technology again. The dictum for the eighteenth century that science owed more to technology than *vice versa* was temporarily inapplicable in the nineteenth century. The relations between science and technology, if they existed at all, were unsystematic for the reasons given above in the characterization of paradigmatic science. One new pattern that emerged was the application of science to technology. This process is essentially subject to chance as the production of knowledge in the context of academic scientific research was, in principle, not oriented to technological uses and it was only in the context of the latter that the relevance of scientific knowledge for practical uses was determined. Typically the process of application demanded substantial efforts of translation. With the beginning of the twentieth century the relation of science and technology gradually became closer. While in some areas, such as those related to prime movers, the engineering work could rest on theoretical foundations that had been provided by science a century and more earlier, in others technological advances rested on contemporary or very recent scientific discoveries, as, for example, in the case of the radio (Marconi, 1895–1907) and the Fleming valve (1904) which were closely connected with Maxwell's work on electromagnetism (1873) (44) Another example is chemistry where the development of the periodic table (Mendeleev, 1871) and the development of the basic theories of chemical compounds (Kekulé, 1858–1866) prepared the ground for what was to become the first 'science-based-industry'. Yet, a pattern quite different from this had already became apparent, for example with Redtenbacher's attempt to formulate the laws of locomotive construction. This rather early development may be taken to be either the induction of a special theory development into science or the transfer of theory construction into practical goals, depending on the point of reference one assumes. The reason for that to become possible is to be found in the stage of development of science, i.e., in the degree of maturity or perfection the disciplines have reached, or in other words, in the accumulated body of theoretical knowledge.

It is a matter of philosophical debate whether, given the dynamics of 'normal research', the development of science will eventually reach a point where all general laws and principles of nature are known and scientific theories are therefore 'closed'. Heisenberg has made this claim by assuming that the laws contained in closed theories will always prove valid and von Weizsäcker has done likewise (45). Classical mechanics, dynamics, the special theory of relativity, and quantum mechanics are given as examples. Even if one remains sceptical about such a far reaching assertion, it is evident that the pattern of development of these firmly established theories, like that of the natural sciences as a whole, differs in some important respects from that in the preceeding phases: the 'revolutions' that occur, if any, do not overthrow the entire theoretical knowledge accumulated before, but rather redefine its range of validity, there are fewer, if any, opposed schools that represent competing views and theories on the subject matter or set of problems, and finally, the occurrence of far reaching fundamental contributions to theory and of spectacular generalizations has decreased. Among scientists themselves there seems to be a growing conviction that some problems have been solved once and for all, like those of motion, heat, light and electricity, since quantum theory has explained in principle the structure of the atom and of matter under terrestrial conditions (v. Weisskopf) and that of heredity has been solved by molecular genetics (G. Stent) (46).

This view is further underscored by the National Academy's evaluation of eight subfields of physics, out of which only 'Elementary Particles' and 'Astrophysics and Relativity' are granted a definite 'potential for discovery of fundamental laws' while the others (e.g., Acoustics, Optics, Plasma and Fluids, etc.) are credited with some "potential for discovery of generalization of broad scientific applicability", ranging high in their "potential contribution to technology" (47).

Obviously, such an advanced stage of development, which for conceptual convenience we may call 'postparadigmatic', has implications for the process of knowledge production. Harvey Brooks in elaborating Weisskopf's distinction between 'intensive' and 'extensive' research has described them unwittingly when he writes: "It (extensive research-P.W.) aims at elucidating the applicability of fairly well-understood principles and theories to an increasing variety of systems, often of increasing complexity . . . Most of chemistry, solid-state physics, and systematic biology are examples of extensive research areas, in which the fundamental principles are understood and the task of research is to discover precisely how they apply to real objects or systems. Extensive research is more likely to be related to applications than is intensive research, and it is no accident therefore, that most of even the most basic industrial research is of the extensive variety. Indeed, *in extensive research the possible ramifications of the underlying principles are usually so diverse and varied that considerations of possible applicability are almost necessary to assist in the selection of problems and research directions*" (48).

If Brook's description is correct, the essential new element in those areas of research which have reached the extensive or 'post-paradigmatic' stage is that the internal set of criteria of relevance that guided the process of 'normal science' and was the crucial prerequisite of the autonomy of paradigmatic science from competing social criteria of relevance loses its directive influence over the process of problem selection and, thus, knowledge production. From this is has been concluded that an "orientation of theory development to external goals becomes possible" or, that "the internal regulatives of development can be replaced by external ones or at least are compatible with them" (49).

If we look at the development of technological research, some characteristics can be identified which, in a particular way, add to those of the scientific development. One is the increasingly frequent situation that the technical research effort, because of the very nature of the problems dealt with is carried over into the realm of basic scientific research. Thus, studies of the behavior of metals have led to crystallographical analyses of a much more basic nature and the technical approaches to achieve controlled nuclear fusion as a source of energy have run into difficulties because the nature of plasmas was not well enough understood and consequently more basic research was required (50). In all these cases, Rapp points out, the technical problems lead to further basic analyses in the natural sciences and consequently the direction of their development is *also* determined by the technical problems.

Another characteristic is the development of specific theories for technical phenomena. One among several examples is the case of chemical engineering (science) which emerged as a purely empirical-experimental procedure and was successively theorized until the representation of the real processes in chemical reactors became possible with the means of thermodynamics, hydrodynamics and kinetics (51). From the viewpoint of technology such a 'theorization' means that basic operations can be traced back to more general theories. From the viewpoint of science it means that general scientific theories are made concrete for specific phenomena (52). In cognitive terms, then, wherever this pattern prevails, the differentiation between science and technology becomes blurred if not altogether suspended. Again, it would be misleading to speak of an influence in only one direction. Rather, the new pattern emerges because two conditions are mutually supporting each other: the development of scientific knowledge has reached a point where its explanatory and predictive power can be extended to a rapidly increasing variety of phenomena; technical problems have reached such a complexity that their solution requires the use of scientific methods, notably theory construction based on mathematical description and systematic experiments. Essentially, this amounts to a subjection of practical (i.e., economic, political, social) criteria of relevance to those of truth or, in other words, the extension of the latter to principally all areas of practice.

Some novel types of institutions support the impression conveyed by the changes in the cognitive configurations of science and technology. A very early example for the institutionalization of applied research, a function that appears to be a logical consequence of the separation of science and technology, wherever it occurred, is the foundation of the 'Physikalisch-Technische Reichsanstalt'. Independent from concerns of immediate utilization and free from the teaching function it was supposed to organize the transfer of scientific findings to technological applications. Divided into a technical-mechanical and a physical-scientific department the PTR was devoted to both basic research and applied research and technology. Incidentally, its foundation was a reaction to the needs of improvement of scientific precision instruments, materials testing and measuring which had not kept pace with industrial practice. Its task was defined along two lines: the physics department was to carry out physical analyses and measurements which were primarily aimed at the solution of scientific problems of great importance in a 'scientific or technical direction' (53), the technical department was supposed to gauge measures and weights, to determine material constants and extrapolate the results of basic research on the technological side. Regardless of its later development this was the institutionalization of the mediating function between science and technology which found many successors elsewhere.

This type of institution is superseded by, or developed further into, yet another one reflecting the even closer connection between science and technology as described above. One example is the laboratory in the 'science-based-industries' (especially electronics, chemistry, communication and instruments) with its interdisciplinary organization of research groups, a supply of highly trained personnel as a criterion of location, the scope of technological development programs and other characteristics which move these industries very near to the organization of science itself (54). Nelson *et al.* point out that also a new strategy of support has appeared in that emphasis is put on "certain key areas where improved understanding is judged particularly likely to yield solutions to practical problems and open up promising areas for development". In the 'science based industries', therefore, an important differentiating criterion between science and technology is disappearing: "Research aimed at opening up new possibilities has substituted both for chance development in the relevant sciences, and for the classical major inventive effort aimed at cracking open a problem through direct attack" (55).

Another, final example is to be seen in the emergence of specific transfer institutions which we have called 'hybrid communities' elsewhere (56). They are organizational frameworks, in which scientists, politicians, administrators and representatives of industry and other interest groups communicate directly in order to determine problem definitions, research strategies and solutions. This involves a complex translation process of political goals into technological objectives and research strategies as disparate universes of discourse are involved. As of now no unified institutional setting for this particular pattern of communication has emerged. The variety of organizational structure is great, ranging from government agencies with in-house research and evaluation capacities such as the NIH, to expert advisory bodies in which government officials, scientists, and practitioners are represented such as the German Council of Education. Their function, however, is invariably the processing of the increased communication load resulting from the differentiation of the political, technological and scientific systems. This institutional development as it is revealed by these different types of organizations is a consequence of the separation of the production of theoretical knowledge and of practical applications of that knowledge. Being separated institutionally the problem becomes the mediation of these different social contexts, a problem typically occurring with differentiation and leading to further complexity. This is also the essence of the process of scientification, but the problem remains why the context of the production of theoretical knowledge does not simply become independent of the various other social contexts as is in fact a tendency. This problem may be answered by looking at the macroanalytical level of the social production and diffusion of knowledge again.

The development we have described in this section may be termed the 'scientification' and subsequent reflexivity of technology and social organization as a whole. With 'scientification' we mean that scientific knowledge increasingly determines social processes. First of all it is applied to technology but beyond that economic development, specialization and division of labour as well as the organization, direction and control of social subsystems are made possible through the application of systematic knowledge. In a sense this means that social practice is linked up to the development of science. Wherever it is feasible, scientific knowledge is transferred into practical contexts. In the early stages of this phase the process of scientific development, i.e., disciplinary theory construction, remains, in principle, oriented to internal criteria of relevance while parallel to that, institutional provisions for application and development of scientific knowledge emerge and expand. One can say that once science has been established it increasingly demonstrated its superiority as a medium to organize and control experience. The extention of the utilization of scientific knowledge into other contexts of social practice, therefore, is an 'inevitable' evolutionary step. It is 'inevitable' because the scientification is realized, so to speak, *through* other relevance criteria rather than against them.

There is the further complementary aspect of this development: reflexivity. Given the availability and utilization of scientific knowledge in other contexts of social practice these become increasingly perceived in terms of scientific knowledge or rather in terms of their accessibility to 'scientific knowing', i.e., systematic explanation, reliable forecasting, etc. One mechanism in the dynamics of scientific development and the evolution of social practice is that increasingly problems become the object of problem-solving actions only after their causes have been determined manipulable. As such, knowledge depends on prior achievements in science. The second mechanism is that as social practice becomes reflexive it generates problems the accessibility of which to 'scientific knowing' is a part of that practice and which, consequently, become the object of systematic forms of experience. This mechanism becomes evident in the anticipatory reference to and reliance on the *potential* of science for the control of *future* secondary consequences of present decisions which is increasingly characteristic of political action (57). Decision-making, in other words, is no longer based alone on *available* knowledge but with *anticipated* secondary consequences in mind. As these cannot be known through experience, or if so only at very high social costs they require forecasting based on systematic knowledge about the objects concerned. That entails a growing 'pressure' from all areas of social practice on the production of knowledge. This is no longer restricted to academic science in the institutionalized sense but reaches into other areas of social practice. From the viewpoint of institutionalized science this appears as goal or problem-orientation. The stage of reflexivity is thus characterized by an interdependence of the scientific reflexion of 'practical' goals and a goal orientation of scientific development. It is in the 'scientification' of technology where this configuration becomes first visible. But there are obvious indications in the structure and social organization of the production of knowledge and learning.

The scientification and reflexivity of the political and economic sectors are, of course, reflected in the expansion of the administrative and planning bodies. With respect to the orientation of research to practical goals we have mentioned already the transfer agencies and 'hybrid communities'. Perhaps more revealing yet is the growing volume of 'secondary disciplines', e.g., energy research, environmental research, policy research and many others. These indicate, among other things, that problem areas of public policy have been turned into subject matters of scientific research or, also that science is becoming oriented to external goals. Clearly, the establishment of these fields demarcates the advent of a new, 'horizontal' structure of science. The extension of the scientific method to other areas of social practice, as we have described it, thus transcends the contemporary forms of institutionalized science as they have emerged in the nineteenth century.

The same development can be traced in the education sector, i.e., with respect to the diffusion of knowledge. Here, the disparity between academic-theoretical and occupational practical training has become a major problem of concern, indicating the degree to which the system has become both cognitively and institutionally autonomous in the past. As it is realized that the demands of the occupational structure cannot fully determine the contents of educational processes, the only criterion of their organization left is that they are responsive to diverse social problem situations.

The entire complex of efforts that is reflected in concepts such as 'interdisciplinarity', 'didactics', 'curriculum construction' can be summarily looked upon as an attempt to integrate the disparate spheres of knowledge and practice by means of a (meta-)reflexion of social and political goals and the orientation of the production and diffusion of knowledge to problems of practice in order to achieve a scientifically guided practice. This constitutes the reflexivity of learning processes which, as is evident in the extension and elaboration of the curriculum concept, amounts to an intellectual reconstruction of reality. This scientific endeavor is no longer limited to the sphere of production but has been extended to practically all other areas of human life, among them such seemingly 'natural' activities as eating and love. Experience as a regulative of action is, thus, further replaced by systematic knowledge and the necessity of formalized processes of learning extended to a growing number of areas of action.

As the range of knowledge that can be 'learned' is rapidly expanding and the spheres of action are continuously restructured, a fundamental consequence for the structure of learning arises. The adaption to the constant change of conditions through experience becomes less and less possible, and consequently the biographical division of learning and the application of knowledge loses its functionality. We presently witness the first indication that the generation change as the chief mechanism determining the quantity and contents of the knowledge to be learned becomes supplemented (and eventually replaced?) by the principle of life long learning which, with respect to the organization of the diffusion of knowledge becomes the characteristic element of the 'reflexivity' of social organization. The diffusion of systematic knowledge, and the adaptation to its growth tends to become a continuous process both inter- and intra-biographically.

Notes and References

1. cf. S. C. Gilfillan, *The Sociology of Invention*, Cambridge, Mass.: MIT Press, 1970, for one of the few genuine sociological studies of technology.
2. cf. R. K. Merton, *Science, Technology and Society in Seventeenth-Century England*, New York: Harper & Row, 1970, Chapters VII–IX, also preface, 1970, XIII.
3. cf. Gilfillan, *op. cit.*, 1970, Note 1.
4. cf. R. Nelson 'The Simple Economics of Basic Scientific Research', in N. Rosenberg (ed.), *The Economics of Technological Change*, Middlesex: Penguin, 1971, pp. 148–163.
5. cf. D. de Solla Price, 'The Structures of Publication in Science and Technology', in W. H. Gruber and D. G. Marquis (eds.), *Factors in the Transfer of Technology*, Cambridge: MIT Press, 1969, p. 99.
6. P. Mathias, 'Preface', in A. E. Musson (ed.), *Science, Technology and Economic Growth in the Eighteenth Century*, London: Methuen, 1972. Cf. also: A. E. Musson, 'Editor's Introduction', in *ibid*.
7. For this 'three-phase-model' of scientific development which is an extension of Kuhn's notion of pre-paradigmatic and paradigmatic science, I follow G. Böhme *et al.* (in this volume) except for some minor alterations. The model is a simplification insofar as the different disciplines go through theses stages at different times. With respect to the traditional periodization of these stages only physics and to some extent chemistry fit into this pattern.
8. Cf. the projects 'Hindsight' and 'Traces'.
9. P. Mathias, 'Preface', *op. cit.*, Note 6.
10. cf. the similar notion of 'images of knowledge' in: Y. Elkana, 'Rationality and Scientific Change', (unpubl. ms.).
11. cf. W. Krohn, 'Die 'Neue Wissenschaft' der Renaissance', in G Böhme *et al*. *Experimentelle Philosophie*, Frankfurt: Suhrkamp, 1977, pp. 13–128, p. 45. cf. also E. Zilsel, 'The Genesis of the Concept of Scientific Progress', *Journal for the History of Ideas* 6, 1945, pp. 334, 336.
12. cf. E. Zilsel, 'The Sociological Roots of Science', *American Journal of Sociology* 47, 1942, pp 245–279; cf. also G. de Santillana, 'The Role of Art in the Scientific Renaissance', in M. Clagett (ed.), *Critical Problems in the History of Science*, Madison: University of Wisconsin Press, 1959, pp. 33–65.
13. Elkana, *op. cit.*, p. 82.
14. *ibid.*, p. 84.
15. F. Sprat, *History of the Royal Society*, London: Routledge & Kegan Paul, 1966, p. 83.
16. P. Rossi, *Philosophy, Technology and the Arts in the Early Modern Era*, New York: Harper & Row, 1970, p. 149.
17. *ibid.*, p. 161.
18. R. Descartes, *Von der Methode*, Hamburg: Meiner Verlag, 1960, p. 50.
19. G. Böhme, *et al.*, in this volume.
20. cf. M. Ornstein, *The Role of Scientific Society in the Seventeenth Century*, Chicago, 1928, p. 53.
21. *ibid.*, p. 90.

22. cf. Essays of Natural Experiments Made in the Academie del Cimento, New York/London (1648): Johnson Reprint Corp. 1964.
23. M. Ornstein, *op. cit.*, pp. 91 and 104, Note 20.
24. Cited in Ornstein, *op. cit.*, pp. 104 and 108, my italics.
25. F. Sprat as cited in A.R.J.P. Ubbelode, 'The Beginnings of the Change from Craft Ministry to Science as a Basis for Technology', in C. Sinclair *et al.* (eds.) *A History of Technology*, vol. IV, Oxford: University Press, 1958.
26. cf. R. K. Merton, *op. cit.*, p. 173, Note 2.
27. cf. R. Hahn, *The Anatomy of a Scientific Institution – The Paris Academy of Sciences 1666–1803*, Berkeley *et al.* 1971, p. 14.
28. *ibid.*, p. 24.
29. cf. L. Stone 'The Size and Composition of the Oxford Student Body 1580–1910', in L. Stone (ed.) *The University in Society*, Princeton: Princeton University Press, 1974, p. 72.
30. T. Kuhn, The Structure of Scientific Revolutions, Chicago: Chicago University Press, 2nd edn., 1970, p. 23.
31. *ibid.*, 25 pp.
32. Böhme *et al.*, *op. cit.* cf. there for further discussion and examples, Note 7.
33. cf. *ibid.*
34. cf. N. Luhmann, 'Selbststeuerung der Wissenschaft', in N. Luhmann, *Soziologische Aufklärung*, Köln/Opladen: Westdeutscher Verlag, 1970, 234 pp.
35. H. Schelsky, *Einsamkeit und Freiheit*, Düsseldorf: Bertelsmann Universitätsverlag, 1971, 38 pp.
36. R. Stichweh, *Ausdifferenzierung der Wissenschaft – Eine Analyse am deutschen Beispiel*, Science Studies Report No. 8, Universität Bielefeld, 1977, p. 80.
37. cf. J. Liebig, *Über das Studium der Naturwissenschaften und über den Zustand der Chemie in Preußen*, Braunschweig, 1940, p. 38.
38. cf. Stichweh, *op. cit.*, 178 pp., Note 36.
39. "Institutionalization was also taking place outside the schools in a variety of ways, and while institutionalization in engineering colleges of one sort or another was very important, it was by no means the whole story. The emphasis on schools is, so far as engineering professionalization is concerned, to some extent a scholarly convenience. The history of engineering professionalism in these years has yet to be written, and those who study the subject, must assemble rather scattered data from many sources. The only part of this story (professionalization) that is even partially available in reasonable form is that dealing with schools. But professionalization also was taking place outside the schools (e.g., the British case is extreme, schooling came in very late, long after professionalization had already taken place, notably around the figure of John Smeaton, who developed or adapted experimental methods from science, for engineering and whose followers formed the 'Smeatonians' an early engineering professional society or a proto-type of the ones that were founded shortly there-after). Besides professional societies like the Institution of Civil Engineers, the scientific academies were important (for example, as an outlet for the publication of scientific papers by engineers such as that by Borda; also through the system of refereeing papers submitted and printing reports). Another factor were national (governmental) organizations, such as the Corps du Génie, founded by Vauban. Clearly, the first engineering school, the École des Ponts et Chaussees was an outgrowth of such a government agency, the Corps

des Ponts et Chaussees, and for many years it was called the 'central office' and only later got the designation 'École'." Communication of E. Layton to the author. June 13, 1977. Also, a PAREX study group is concerned with some aspects of the national differences with respect to the institutional frameworks for science and engineering teaching and research and I thank P. Lundgreen for drawing my attention to that.

40. cf. P. Lundgreen, *Techniker in Preußen während der frühen Industrialisierung*, Berlin: Colloquium Verlag, 1975, and Peter Lundgreen, Bildung und Wirtschaftswachstum im Industrialisierungsprozeß des 19. Jahrhunderts, Berlin: Colloquium Verlag, 1973, 134 pp.

41. cf. K. H. Manegold, *Universität, Technische Hochschule und Industrie*, Berlin: 1970, 39 pp.

42. cf. K. H. Manegold, 'Das Verhältnis von Naturwissenschaft und Technik im 19. Jahrhundert im Spiegel der Wissenschaftsorganisation', in *Verein Deutscher Ingenieure, Technikgeschichte in Einzeldarstellungen*, Nr. 11, Düsseldorf: VDI-Verlag, 1969, 165 pp.

43. Riedler, cited in Manegold, *Universität, Technische Hochshule und Industrie op. cit.* 153, Note 41. cf. also Manegold, 'Das Verhältnis von Naturwissenschaft und Technik', *op. cit.* p. 117, Note 42.

44. cf. H. Bode, 'Reflections on the Relation Between Science and Technology', in National Academy of Sciences (ed.) *Basic Research and National Goals*. A Report to the Committee on Science and Astronautics, U.S. House of Representatives, Washington D.C.: U.S. Government Printing Office, 1965, p. 48.

45. cf. W. Heisenberg, 'Der Begriff abgeschlossene Theorie', in W. Heisenberg, *Schritte über Grenzen*, München: Piper Verlag, 1971, p. 93; C.F.v. Weizsäcker, *Die Einheit der Natur*, München: Hanser Verlag, 1971, p. 193.

46. cf. W. van den Daele, 'Scientific Development and External Goals', Paper presented at the 'International Seminar on Science Studies', The Academy of Finland, 11–14 January, 1977 for this and part of the following argument.

47. National Academy of Sciences (ed.), *Physics in Perspective*, Washington: Printing and Publishing Office, NAS, 1973, p. 404.

48. H. Brooks, 'Applied Research, Definitions, Concepts, Themes', in National Academy of Sciences (ed.), *Applied Science and Technological Progress*, A Report to the Committee on Science and Astronautics, U.S. House of Representatives, Washington D.C.: U.S. Government Printing Office, 1967, , 26, my italics.

49. W. van den Daele, *op. cit.*, p. 16, Note 46.

50. cf. National Academy of Sciences, *op. cit.*, p. 221, Note 47; also G. Küppers; 'Die geplante Wissenschaft, – Die Fusionsforschung als Beispiel für die Steuerbarkeit in der Grundlagenforschung', unpublished ms. USP-Wissenschaftsforschung, Bielefeld, 1976; F. Rapp, 'Technology and Natural Science – A Methodological Investigation'. in F. Rapp (ed.), *Contributions to a Philosophy of Technology*, Dordrecht: D. Reidel, 1974.

51. K. Buchholz, 'Zu Stand und Entwicklung der Verfahrenstechnik', unpublished ms., Frankfurt/M., 1974.

52. cf. Böhme *et al.*, *op. cit.*, Note 19, also there for discussion of further examples.

53. Cited in Manegold, Das Verhältnis von Naturwissenschaft und Technik, *op. cit.*, p. 182, Note 42.

54. cf. H. Bode, *op. cit.*, 62 pp., Note 44.

55. R. Nelson, *op. cit.*, 42f, Note 4.
56. cf. for a more detailed analysis of these institutions W. van den Daele, W. Krohn, P. Weingart, 'The Political Direction of Scientific Development', in E. Mendelsohn, P. Weingart, R. Whitley (eds.), *The Social Production of Scientific Knowledge*, Sociology of the Sciences, A Yearbook vol. I, Dordrecht: D. Reidel 1977, pp. 219-242.
57. cf. the Energy debate and the anticipated solution of the fusion and nuclear waste disposal problems on which policy is based, cf. W. van den Daele, P. Weingart, 'Resistance and Receptivity of Science to External Direction: the Emergence of New Disciplines under the Impact of Science Policy', in G. Lémaine *et al.* (eds.), *Perspectives on the Emergence of Scientific Disciplines*, The Hague, Paris: Mouton, 1976.

INDEX

287